Teacher Edition

Eureka Math
Grade 5
Module 3

Special thanks go to the Gordon A. Cain Center and to the Department of Mathematics at Louisiana State University for their support in the development of *Eureka Math*.

Published by the non-profit Great Minds

Printed in the U.S.A.
This book may be purchased from the publisher at eureka-math.org
10 9 8 7 6 5 4 3

ISBN 978-1-63255-380-5

***Eureka Math: A Story of Units* Contributors**

Katrina Abdussalaam, Curriculum Writer
Tiah Alphonso, Program Manager—Curriculum Production
Kelly Alsup, Lead Writer / Editor, Grade 4
Catriona Anderson, Program Manager—Implementation Support
Debbie Andorka-Aceves, Curriculum Writer
Eric Angel, Curriculum Writer
Leslie Arceneaux, Lead Writer / Editor, Grade 5
Kate McGill Austin, Lead Writer / Editor, Grades PreK–K
Adam Baker, Lead Writer / Editor, Grade 5
Scott Baldridge, Lead Mathematician and Lead Curriculum Writer
Beth Barnes, Curriculum Writer
Bonnie Bergstresser, Math Auditor
Bill Davidson, Fluency Specialist
Jill Diniz, Program Director
Nancy Diorio, Curriculum Writer
Nancy Doorey, Assessment Advisor
Lacy Endo-Peery, Lead Writer / Editor, Grades PreK–K
Ana Estela, Curriculum Writer
Lessa Faltermann, Math Auditor
Janice Fan, Curriculum Writer
Ellen Fort, Math Auditor
Peggy Golden, Curriculum Writer
Maria Gomes, Pre-Kindergarten Practitioner
Pam Goodner, Curriculum Writer
Greg Gorman, Curriculum Writer
Melanie Gutierrez, Curriculum Writer
Bob Hollister, Math Auditor
Kelley Isinger, Curriculum Writer
Nuhad Jamal, Curriculum Writer
Mary Jones, Lead Writer / Editor, Grade 4
Halle Kananak, Curriculum Writer
Susan Lee, Lead Writer / Editor, Grade 3
Jennifer Loftin, Program Manager—Professional Development
Soo Jin Lu, Curriculum Writer
Nell McAnelly, Project Director

Ben McCarty, Lead Mathematician / Editor, PreK–5
Stacie McClintock, Document Production Manager
Cristina Metcalf, Lead Writer / Editor, Grade 3
Susan Midlarsky, Curriculum Writer
Pat Mohr, Curriculum Writer
Sarah Oyler, Document Coordinator
Victoria Peacock, Curriculum Writer
Jenny Petrosino, Curriculum Writer
Terrie Poehl, Math Auditor
Robin Ramos, Lead Curriculum Writer / Editor, PreK–5
Kristen Riedel, Math Audit Team Lead
Cecilia Rudzitis, Curriculum Writer
Tricia Salerno, Curriculum Writer
Chris Sarlo, Curriculum Writer
Ann Rose Sentoro, Curriculum Writer
Colleen Sheeron, Lead Writer / Editor, Grade 2
Gail Smith, Curriculum Writer
Shelley Snow, Curriculum Writer
Robyn Sorenson, Math Auditor
Kelly Spinks, Curriculum Writer
Marianne Strayton, Lead Writer / Editor, Grade 1
Theresa Streeter, Math Auditor
Lily Talcott, Curriculum Writer
Kevin Tougher, Curriculum Writer
Saffron VanGalder, Lead Writer / Editor, Grade 3
Lisa Watts-Lawton, Lead Writer / Editor, Grade 2
Erin Wheeler, Curriculum Writer
MaryJo Wieland, Curriculum Writer
Allison Witcraft, Math Auditor
Jessa Woods, Curriculum Writer
Hae Jung Yang, Lead Writer / Editor, Grade 1

This page intentionally left blank

Mathematics Curriculum

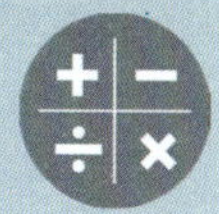

Table of Contents

GRADE 5 • MODULE 3

Addition and Subtraction of Fractions

Grade 5 • Module 3

Addition and Subtraction of Fractions

OVERVIEW

In Module 3, students' understanding of addition and subtraction of fractions extends from earlier work with fraction equivalence and decimals. This module marks a significant shift away from the elementary grades' centrality of base ten units to the study and use of the full set of fractional units from Grade 5 forward, especially as applied to algebra.

In Topic A, students revisit the foundational Grade 4 standards addressing equivalence. When equivalent, fractions represent the same amount of area of a rectangle and the same point on the number line. These equivalencies can also be represented symbolically.

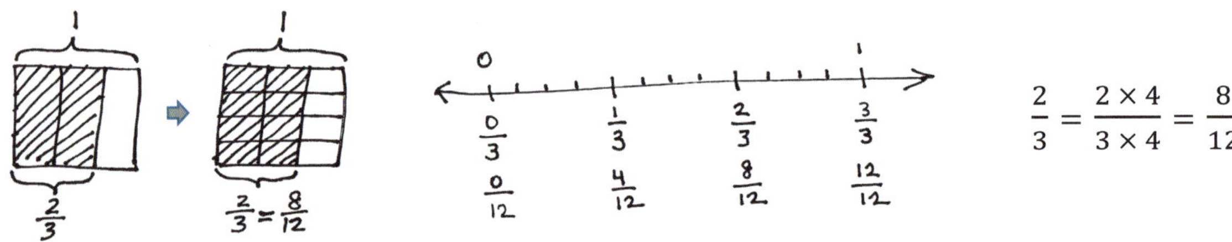

$$\frac{2}{3} = \frac{2 \times 4}{3 \times 4} = \frac{8}{12}$$

Furthermore, equivalence is evidenced when adding fractions with the same denominator. The sum may be decomposed into parts (or recomposed into an equal sum). An example is shown as follows:

$$\frac{2}{3} = \frac{1}{3} + \frac{1}{3}$$

$$\frac{7}{8} = \frac{3}{8} + \frac{3}{8} + \frac{1}{8}$$

$$\frac{6}{2} = \frac{2}{2} + \frac{2}{2} + \frac{2}{2} = 1 + 1 + 1 = 3$$

$$\frac{8}{5} = \frac{5}{5} + \frac{3}{5} = 1\frac{3}{5}$$

$$\frac{7}{3} = \frac{6}{3} + \frac{1}{3} = 2 \times \frac{3}{3} + \frac{1}{3} = 2 + \frac{1}{3} = 2\frac{1}{3}$$

This also carries forward work with decimal place value from Modules 1 and 2, confirming that like units can be composed and decomposed.

5 tenths + 7 tenths = 12 tenths = 1 and 2 tenths

5 eighths + 7 eighths = 12 eighths = 1 and 4 eighths

In Topic B, students move forward to see that fraction addition and subtraction are analogous to whole number addition and subtraction. Students add and subtract fractions with unlike denominators (**5.NF.1**) by replacing different fractional units with an equivalent fraction or like unit.

1 fourth + 2 thirds = 3 twelfths + 8 twelfths = 11 twelfths

$$\frac{1}{4}+\frac{2}{3}=\frac{3}{12}+\frac{8}{12}=\frac{11}{12}$$

This is not a new concept, but certainly a new level of complexity. Students have added equivalent or like units since kindergarten, adding frogs to frogs, ones to ones, tens to tens, etc.

1 boy + 2 girls = 1 child + 2 children = 3 children

1 liter – 375 mL = 1,000 mL – 375 mL = 625 mL

Throughout the module, a concrete to pictorial to abstract approach is used to convey this simple concept. Topic A uses paper strips and number line diagrams to clearly show equivalence. After a brief concrete experience with folding paper, Topic B primarily uses the rectangular fractional model because it is useful for creating smaller like units by means of partitioning (e.g., thirds and fourths are changed to twelfths to create equivalent fractions as in the diagram below). In Topic C, students move away from the pictorial altogether as they are empowered to write equations clarified by the model.

$$\frac{1}{4}+\frac{2}{3}=\left(\frac{1\times 3}{4\times 3}\right)+\left(\frac{2\times 4}{3\times 4}\right)=\frac{3}{12}+\frac{8}{12}=\frac{11}{12}$$

Topic C also uses the number line when adding and subtracting fractions greater than or equal to 1 so that students begin to see and manipulate fractions in relation to larger whole numbers and to each other. The number line allows students to pictorially represent larger whole numbers. For example, "Between which two whole numbers does the sum of $1\frac{3}{4}$ and $5\frac{3}{5}$ lie?"

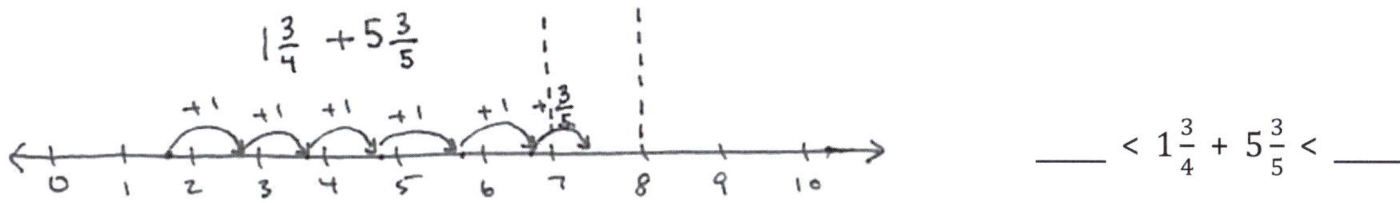

$\_\_\_\_ < 1\frac{3}{4} + 5\frac{3}{5} < \_\_\_\_$

This leads to an understanding of and skill with solving more complex problems, which are often embedded within multi-step word problems:

> *Cristina and Matt's goal is to collect a total of* $3\frac{1}{2}$ *gallons of sap from the maple trees. Cristina collected* $1\frac{3}{4}$ *gallons. Matt collected* $5\frac{3}{5}$ *gallons. By how much did they beat their goal?*

goal: $3\frac{1}{2}$ gal ?

collected: $1\frac{3}{4}$ gal | $5\frac{3}{5}$ gal

$$1\frac{3}{4} + 5\frac{3}{5} - 3\frac{1}{2} = 3 + \left(\frac{3 \times 5}{4 \times 5}\right) + \left(\frac{3 \times 4}{5 \times 4}\right) - \left(\frac{1 \times 10}{2 \times 10}\right)$$

$$= 3 + \frac{15}{20} + \frac{12}{20} - \frac{10}{20} = 3\frac{17}{20}$$

Cristina and Matt beat their goal by $3\frac{17}{20}$ gallons.

Word problems are a part of every lesson. Students are encouraged to draw tape diagrams, which encourage them to recognize part–whole relationships with fractions that they have seen with whole numbers since Grade 1.

In Topic D, students strategize to solve multi-term problems and more intensely assess the reasonableness of their solutions to equations and word problems with fractional units (**5.NF.2**).

> *"I know my answer makes sense because the total amount of sap they collected is about 7 and a half gallons. Then, when we subtract 3 gallons, that is about 4 and a half. Then, 1 half less than that is about 4.* $3\frac{17}{20}$ *is just a little less than 4."*

The Mid-Module Assessment follows Topic B. The End-of-Module Assessment follows Topic D.

Notes on Pacing for Differentiation

If pacing is a challenge, consider the following modifications and omissions. Omit Lesson 2 as it addresses a Grade 4 standard. In Lesson 3, omit the paper folding exercise, and consider it a remediation tool. Omit the Sprint in Lesson 12, and replace it with simple reasoning about fractions on the number line, such as "Is $\frac{3}{4}$ greater than or less than $\frac{1}{2}$? $\frac{3}{5}$? $\frac{3}{7}$?" In Lesson 15, choose two or three problems, and omit the others. Use the omitted problems as Application Problems in future lessons. Consider omitting Lesson 16 and using it in a center for early finishers, or have advanced students work the problems and present their solutions in a video or interactive demonstration. Consider asking the following questions to students, "Have you ever thought about what the whole would look like if this paper were one-half? What if it were one-third? What if this is three-fourths of the whole? What would the whole look like then?"

Note: In the first year of implementation, beginning in Lesson 5, be sure to include the fluency activities requiring students to subtract fractions less than one from a whole number (e.g., $4 - \frac{5}{8}$) in order to prepare students to subtract larger mixed numbers in Topics B and C. Model these fluency activities on the number line and with a tape diagram.

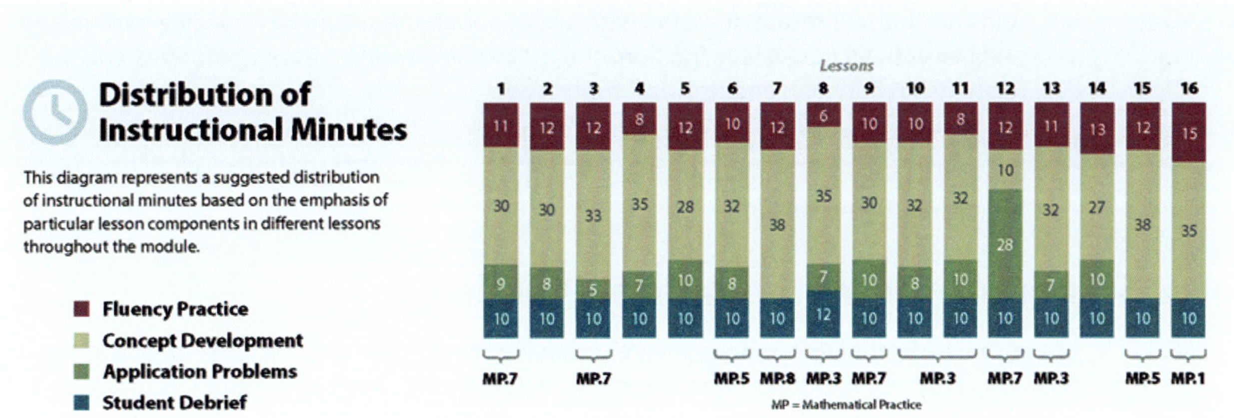

Focus Grade Level Standards

Use equivalent fractions as a strategy to add and subtract fractions.[1]

5.NF.1 Add and subtract fractions with unlike denominators (including mixed numbers) by replacing given fractions with equivalent fractions in such a way as to produce an equivalent sum or difference of fractions with like denominators. *For example, 2/3 + 5/4 = 8/12 + 15/12 = 23/12. (In general, a/b + c/d = (ad + bc)/bd.)*

5.NF.2 Solve word problems involving addition and subtraction of fractions referring to the same whole, including cases of unlike denominators, e.g., by using visual fraction models or equations to represent the problem. Use benchmark fractions and number sense of fractions to estimate mentally and assess the reasonableness of answers. *For example, recognize an incorrect result 2/5 + 1/2 = 3/7, by observing that 3/7 < 1/2.*

Foundational Standards

4.NF.1 Explain why a fraction *a/b* is equivalent to a fraction *(n x a)/(n x b)* by using visual fraction models, with attention to how the number and size of the parts differ even though the two fractions themselves are the same size. Use this principle to recognize and generate equivalent fractions.

4.NF.3 Understand a fraction *a/b* with *a* > 1 as a sum of fractions 1/*b*.

a. Understand addition and subtraction of fractions as joining and separating parts referring to the same whole.

b. Decompose a fraction into a sum of fractions with the same denominator in more than one way, recording each decomposition by an equation. Justify decompositions, e.g., by using a visual fraction model. *Examples: 3/8 = 1/8 + 1/8 + 1/8; 3/8 = 1/8 + 2/8; 2 1/8 = 1 + 1 + 1/8 = 8/8 + 8/8 + 1/8.*

---

[1]Examples in this module also include tenths and hundredths in fraction and decimal form.

c. Add and subtract mixed numbers with like denominators, e.g., by replacing each mixed number with an equivalent fraction, and/or by using properties of operations and the relationship between addition and subtraction.

d. Solve word problems involving addition and subtraction of fractions referring to the same whole and having like denominators, e.g., by using visual fraction models and equations to represent the problem.

Focus Standards for Mathematical Practice

MP.1 **Make sense of problems and persevere in solving them.** Students make sense of problems when they use number lines, tape diagrams, and fraction models to conceptualize and solve fraction addition and subtraction problems. Students also check their work and monitor their progress, assessing their approach and its validity within the given context and altering their method when necessary.

MP.3 **Construct viable arguments and critique the reasoning of others.** As students add and subtract with fractions and mixed numbers, they make choices and reason about which like unit to choose and draw conclusions about what makes some problems simpler than others. Students analyze multiple solution strategies for given problems and draw conclusions about which method is most efficient in each case. Students also critique the reasoning of others and construct viable arguments during this analysis. They also use their understanding of fractions to assess the reasonableness of sums and differences and use these assumptions to justify their conclusions to others.

MP.5 **Use appropriate tools strategically**. Students use mental computation and estimation strategies to assess the reasonableness of their answers. They decide which pictorial model to draw and label and reason about its size relative to the context of the problem. Students decide on the appropriateness of using special strategies when adding and subtracting mixed numbers.

MP.7 **Look for and make use of structure.** Students discern patterns and structures as they draw fraction models and reason about the number of units represented, the size or length of those units, and the name of the fraction that each model represents. They identify patterns in sums and differences when the same fraction is added to or taken from a variety of numbers and use this understanding to generate predictions about the sums and differences.

MP.8 **Look for and express regularity in repeated reasoning.** Students express regularity in repeated reasoning when they look for and use whole number general methods to add and subtract fractions. Adding and subtracting fractions requires finding like units just as it does with whole numbers, such as when adding centimeters and meters.

Overview of Module Topics and Lesson Objectives

Standards		Topics and Objectives		Days
4.NF.1 4.NF.3c 4.NF.3d	A	**Equivalent Fractions**		2
		Lesson 1:	Make equivalent fractions with the number line, the area model, and numbers.	
		Lesson 2:	Make equivalent fractions with sums of fractions with like denominators.	
5.NF.1 **5.NF.2**	B	**Making Like Units Pictorially**		5
		Lesson 3:	Add fractions with unlike units using the strategy of creating equivalent fractions.	
		Lesson 4:	Add fractions with sums between 1 and 2.	
		Lesson 5:	Subtract fractions with unlike units using the strategy of creating equivalent fractions.	
		Lesson 6:	Subtract fractions from numbers between 1 and 2.	
		Lesson 7:	Solve two-step word problems.	
		Mid-Module Assessment: Topics A–B (assessment ½ day, return ½ day, remediation or further applications 2 days)		3
5.NF.1 **5.NF.2**	C	**Making Like Units Numerically**		5
		Lesson 8:	Add fractions to and subtract fractions from whole numbers using equivalence and the number line as strategies.	
		Lesson 9:	Add fractions making like units numerically.	
		Lesson 10:	Add fractions with sums greater than 2.	
		Lesson 11:	Subtract fractions making like units numerically.	
		Lesson 12:	Subtract fractions greater than or equal to 1.	
5.NF.1 **5.NF.2**	D	**Further Applications**		4
		Lesson 13:	Use fraction benchmark numbers to assess reasonableness of addition and subtraction equations.	
		Lesson 14:	Strategize to solve multi-term problems.	
		Lesson 15:	Solve multi-step word problems; assess reasonableness of solutions using benchmark numbers.	
		Lesson 16:	Explore part-to-whole relationships.	

Standards	Topics and Objectives		Days
		End-of-Module Assessment: Topics A–D (assessment ½ day, return ½ day, remediation or further applications 2 days)	3
Total Number of Instructional Days			**22**

Terminology

New or Recently Introduced Terms

- Benchmark fraction (e.g., $\frac{1}{2}$ is a benchmark fraction when comparing $\frac{1}{3}$ and $\frac{3}{5}$)
- Like denominators (e.g., $\frac{1}{8}$ and $\frac{5}{8}$)
- Unlike denominators (e.g., $\frac{1}{8}$ and $\frac{1}{7}$)

Familiar Terms and Symbols[2]

- Between (e.g., $\frac{1}{2}$ is between $\frac{1}{3}$ and $\frac{3}{5}$)
- Denominator (denotes the fractional unit: fifths in 3 fifths, which is abbreviated as the 5 in $\frac{3}{5}$)
- Equivalent fraction (e.g., $\frac{3}{5} = \frac{6}{10}$)
- Fraction (e.g., 3 fifths or $\frac{3}{5}$)
- Fraction greater than or equal to 1 (e.g., $\frac{7}{3}$, $3\frac{1}{2}$, an abbreviation for $3 + \frac{1}{2}$)
- Fraction written in the largest possible unit (e.g., $\frac{3}{6} = \frac{1 \times 3}{2 \times 3} = \frac{1}{2}$ or 1 three out of 2 threes = $\frac{1}{2}$)
- Fractional unit (e.g., the fifth unit in 3 fifths denoted by the denominator 5 in $\frac{3}{5}$)
- Hundredth ($\frac{1}{100}$ or 0.01)
- Kilometer, meter, centimeter, liter, milliliter, kilogram, gram, mile, yard, foot, inch, gallon, quart, pint, cup, pound, ounce, hour, minute, second
- *More than halfway* and *less than halfway*
- Number sentence (e.g., Three plus seven equals ten. Usually written as *3 + 7 = 10.*)
- Numerator (denotes the count of fractional units: 3 in 3 fifths or 3 in $\frac{3}{5}$)
- *One tenth of* (e.g., $\frac{1}{10} \times 250$)
- Tenth ($\frac{1}{10}$ or 0.1)
- Whole unit (e.g., any unit that is partitioned into smaller, equally sized fractional units)
- <, >, =

[2]These are terms and symbols students have seen previously.

Suggested Tools and Representations

- Fraction strips
- Number line (a variety of templates)
- Paper strips (for modeling equivalence)
- Rectangular fraction model
- Tape diagrams

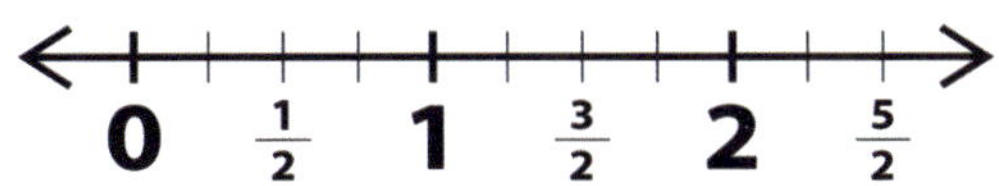

Example of a number line

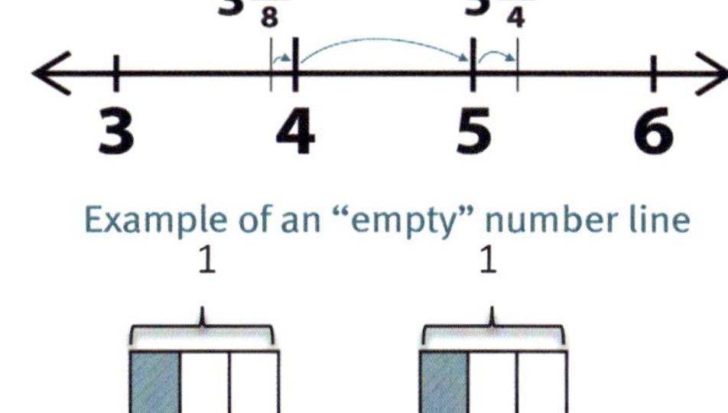

Example of an "empty" number line

Example of a rectangular fraction model

Scaffolds[3]

The scaffolds integrated into *A Story of Units* give alternatives for how students access information as well as express and demonstrate their learning. Strategically placed margin notes are provided within each lesson elaborating on the use of specific scaffolds at applicable times. They address many needs presented by English language learners, students with disabilities, students performing above grade level, and students performing below grade level. Many of the suggestions are organized by Universal Design for Learning (UDL) principles and are applicable to more than one population. To read more about the approach to differentiated instruction in *A Story of Units,* please refer to "How to Implement *A Story of Units*."

Assessment Summary

Assessment Type	Administered	Format	Standards Addressed
Mid-Module Assessment Task	After Topic B	Constructed response with rubric	5.NF.1 5.NF.2
End-of-Module Assessment Task	After Topic D	Constructed response with rubric	5.NF.1 5.NF.2

[3]Students with disabilities may require Braille, large-print, audio, or special digital files. Please visit the website www.p12.nysed.gov/specialed/aim for specific information on how to obtain student materials that satisfy the National Instructional Materials Accessibility Standard (NIMAS) format.

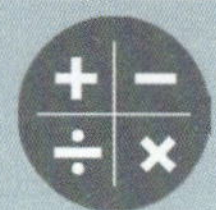

Topic A
Equivalent Fractions

4.NF.1

Focus Standard:	4.NF.1	Explain why a fraction *a/b* is equivalent to a fraction *(n × a)/(n × b)* by using visual fraction models, with attention to how the number and size of the parts differ even though the two fractions themselves are the same size. Use this principle to recognize and generate equivalent fractions.
Instructional Days:	2	
Coherence -Links from:	G4–M5	Fraction Equivalence, Ordering, and Operations
-Links to:	G5–M4	Multiplication and Division of Fractions and Decimal Fractions
	G6–M3	Rational Numbers

In Topic A, students revisit the foundational Grade 4 standards addressing equivalence. When equivalent, fractions can be represented by the same amount of area of a rectangle as well as the same point on a number line. Students subdivide areas and divide number line lengths to model this equivalence. On the number line below, there are 3 × 4 parts of equal length. Both the area model and number line show that 2 thirds is equivalent to 8 twelfths.

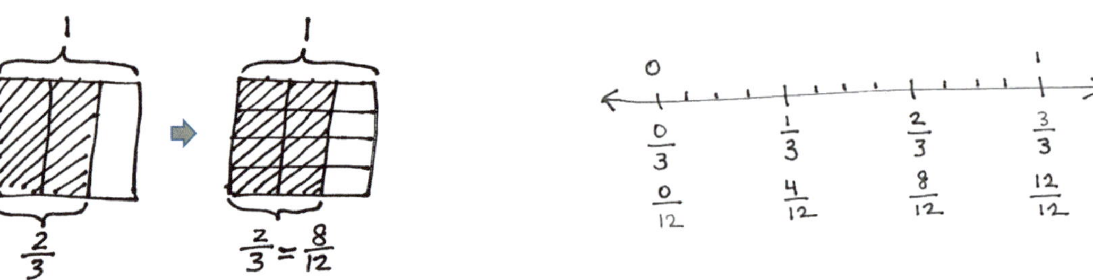

This equivalence can also be represented symbolically as follows:

$$\frac{2}{3} = \frac{2 \times 4}{3 \times 4} = \frac{8}{12}$$

Furthermore, equivalence is evidenced when adding fractions with the same denominator. The sum may be decomposed into parts (or recomposed into an equal sum). An example is shown as follows:

$$\frac{2}{3} = \frac{1}{3} + \frac{1}{3}$$

$$\frac{7}{8} = \frac{3}{8} + \frac{3}{8} + \frac{1}{8}$$

$$\frac{6}{2} = \frac{2}{2} + \frac{2}{2} + \frac{2}{2} = 1 + 1 + 1 = 3$$

$$\frac{8}{5} = \frac{5}{5} + \frac{3}{5} = 1\frac{3}{5}$$

$$\frac{7}{3} = \frac{6}{3} + \frac{1}{3} = 2 \times \frac{3}{3} + \frac{1}{3} = 2 + \frac{1}{3} = 2\frac{1}{3}$$

In Lesson 1, students analyze how and when units must change, particularly when making an equivalent fraction by decomposing larger units into smaller units. This hones their ability to look for and make use of structure (MP.7). They study the area model to make generalizations and then apply those generalizations to work with the number line as they see the same process occurring there within the lengths.

A Teaching Sequence Toward Mastery of Equivalent Fractions	
Objective 1:	**Make equivalent fractions with the number line, the area model, and numbers. (Lesson 1)**
Objective 2:	**Make equivalent fractions with sums of fractions with like denominators. (Lesson 2)**

Lesson 1

Objective: Make equivalent fractions with the number line, the area model, and numbers.

Suggested Lesson Structure

Fluency Practice	(11 minutes)
Application Problem	(9 minutes)
Concept Development	(30 minutes)
Student Debrief	(10 minutes)
Total Time	**(60 minutes)**

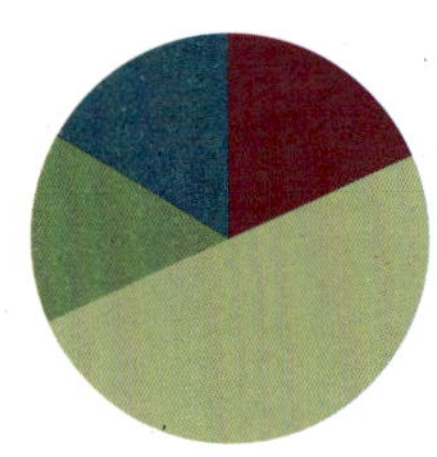

Fluency Practice (11 minutes)

- Sprint: Write the Missing Factor **4.OA.4** (8 minutes)
- Skip-Counting by $\frac{1}{4}$ Hour **5.MD.1** (3 minutes)

Sprint: Write the Missing Factor (8 minutes)

Materials: (S) Write the Missing Factor Sprint

Note: Mentally calculating the missing factor prepares students for making equivalent fractions in today's lesson.

Skip-Counting by $\frac{1}{4}$ Hour (3 minutes)

Note: This fluency activity reviews counting fractions in a real-world context.

T: Let's count by $\frac{1}{4}$ hours. (Rhythmically point up until a change is desired. Show a closed hand, and then point down. Continue, alternating the starting point.)

S: $\frac{1}{4}$ hour, $\frac{2}{4}$ hour, $\frac{3}{4}$ hour, 1 hour, (stop). $1\frac{1}{4}$ hours, $1\frac{2}{4}$ hours, $1\frac{3}{4}$ hours, 2 hours, (stop). $1\frac{3}{4}$ hours, $1\frac{2}{4}$ hours, $1\frac{1}{4}$ hours, (stop).

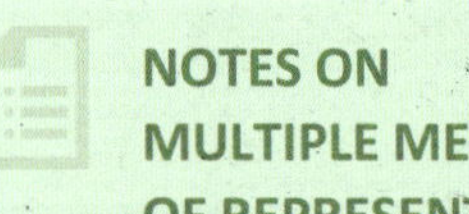

NOTES ON MULTIPLE MEANS OF REPRESENTATION:

Because students are counting units of time, a clock is a natural visual to incorporate. Gesture to points on the clock as students chant, and possibly use it to help signal the direction of the count.

Application Problem (9 minutes)

15 kilograms of rice are separated equally into 4 containers. How many kilograms of rice are in each container? Express your answer as a decimal and as a fraction.

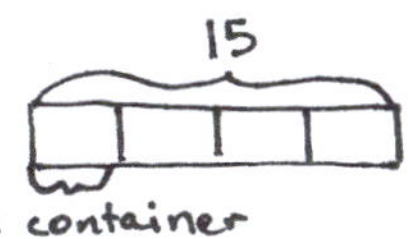

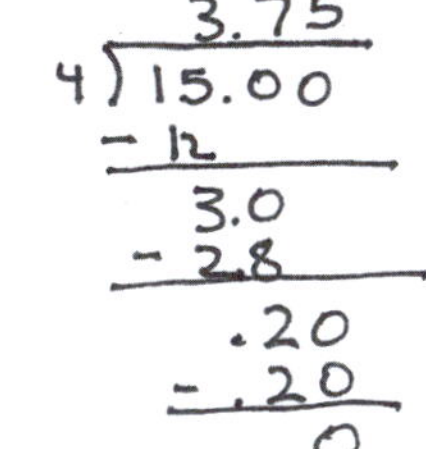

4 units = 15
1 unit = 15 ÷ 4
= 3.75
= $3\frac{75}{100}$

Each container holds 3.75 kg rice.

T: Let's read the problem together.
S: (Read chorally.)
T: Share with your partner: What do you see when you hear the story? What can you draw?
S: (Share with partners.)
T: I'll give you one minute to draw.
T: Explain to your partner what your drawing shows.
T: (After a brief exchange.) What's the total weight of the rice?
S: 15 kilograms.
T: 15 kilograms are being split equally into how many containers?
S: 4 containers.
T: So, the whole is being split into how many units?
S: 4 units.
T: To find 1 container or 1 unit, we have to...?
S: Divide.
T: Tell me the division expression.
S: 15 ÷ 4.
T: Solve the problem on your personal white board. Write your answer both in decimal form and as a whole number and a decimal fraction. (Pause.) Show your board.
T: Turn and explain to your partner how you got the answer. 15 ÷ 4 = 3.75.
T: (After students share.) Show the division equation with both answers.
S: $15 \div 4 = 3.75 = 3\frac{75}{100}$.
T: Express 75 hundredths in its simplest form.
S: 3 fourths.
T: Write your answer as a whole number and a fraction in its simplest form.
S: $15 \div 4 = 3.75 = 3\frac{75}{100} = 3\frac{3}{4}$.
T: So, 3 and 3 fourths equals 3 and 75 hundredths.
T: Tell me your statement containing the answer.
S: Each container holds 3.75 kg or $3\frac{3}{4}$ kg of rice.

Note: This Application Problem reviews division and partitioning as it relates to fractions. Also, it reviews replacing one fraction with another of the same value in anticipation of today's work with equivalent fractions.

Concept Development (30 minutes)

Materials: (S) 4 paper strips $8\frac{1}{2}'' \times 1''$

Problem 1: Make fractions equal to $\frac{1}{2}$.

T: Take a paper strip. Hold it horizontally. Fold it vertically down the middle. How many equal parts do you have in the whole?

S: 2.

T: What fraction of the whole is 1 part?

S: 1 half.

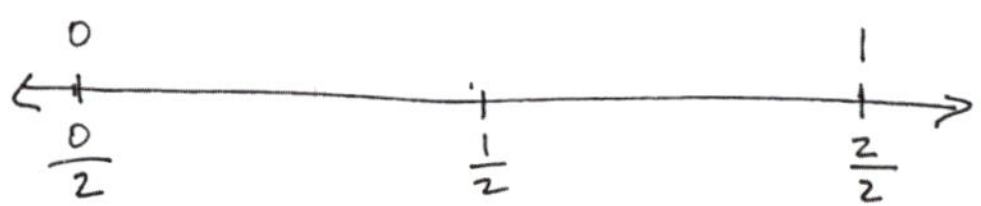

T: Draw a line to show where you folded your paper and label each half, 1 out of 2 units.

T: Make additional paper strips that show thirds, fourths, and fifths.

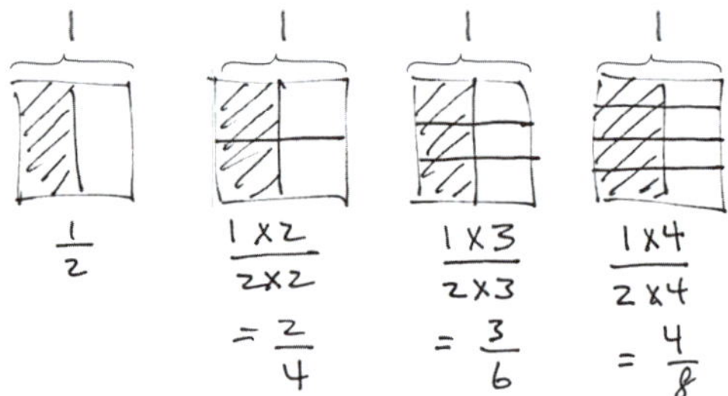

T: (After about 3 minutes.) Draw a number line that is a little longer than your paper strip. Use your strip as a ruler to mark zero and 1 above the line, as well as $\frac{0}{2}$, $\frac{1}{2}$, and $\frac{2}{2}$ below the line.

T: (Allow work time.) Sketch a square that is approximately 1 inch × 1 inch beneath your number line. This is representing the same 1 whole as the number line. For today, show half by vertically dividing the square. Shade 1 half on the left.

T: (Allow work time.) Draw another square to the right of that one. Shade it in the same way to represent $\frac{1}{2}$.

T: Partition 1 half horizontally across the middle.

T: What fraction is shaded now?

S: $\frac{1}{2}$ or $\frac{2}{4}$.

T: (Record numerically, referring to the picture.) 1 group of 2 out of two groups of 2.

$$\frac{1}{2} = \frac{1 \text{ group of two}}{2 \text{ groups of two}} \text{ or } \frac{1 \times 2}{2 \times 2} = \frac{2}{4}$$

T: Explain how we have represented the equivalent fractions to your partners. (Students discuss.)

T: Show me $\frac{2}{4}$ on the number line. (Students show.) Yes, it is exactly the same value as 1 half. It is exactly the same point on the number line.

NOTES ON MULTIPLE MEANS OF ACTION AND EXPRESSION:

Sentence frames help students remember the linguistic and numerical patterns. As they gain confidence, gradually retract the frames. A suitable sentence frame for this lesson would be the equation to the left without any of the digits included.

EUREKA MATH™

T: Work with your partner to draw another congruent square with 1 half shaded. This time, partition it horizontally into 3 equal units (2 lines) and record the equivalent fraction as we did on the first example. If you finish early, continue the pattern.

Problem 2: Make fractions equal to $\frac{1}{3}$.

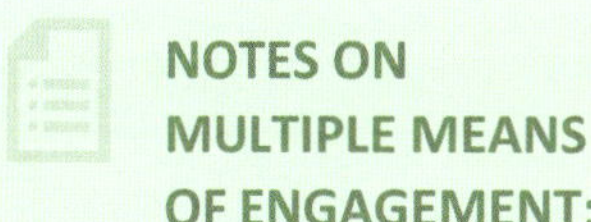

NOTES ON MULTIPLE MEANS OF ENGAGEMENT:

Because this lesson is so pictorial, it is perfect for English language learners. Use hand gestures to support the connection between words and numbers, and numbers and models. For example, say, "one unit of two." (Pause and point to the image and then to the numbers.)

This problem allows students to repeat the procedure with thirds. If necessary for students, try repeating the process thoroughly as outlined in Problem 1. Work with a small group as others work independently, or allow students to work with a partner. It is not necessary for all students to complete the same amount of work. Move on to Problem 3 after about 4 minutes on Problem 2.

Note: Use a model to clarify equivalencies as illustrated below that 6 ninths is shown to be equal to 4 sixths. Go back and ensure that this point is clear with 2 sixths, 3 ninths, and 4 twelfths.

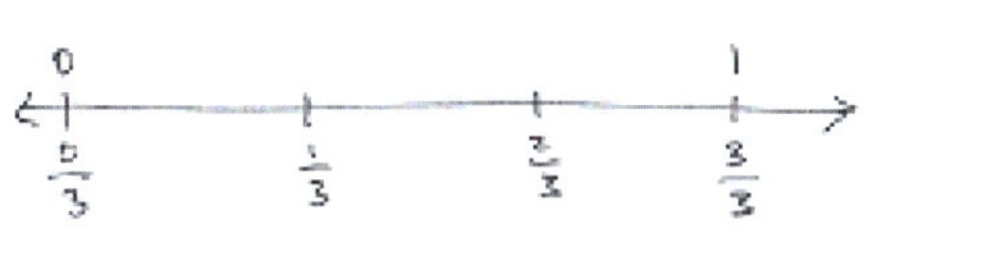

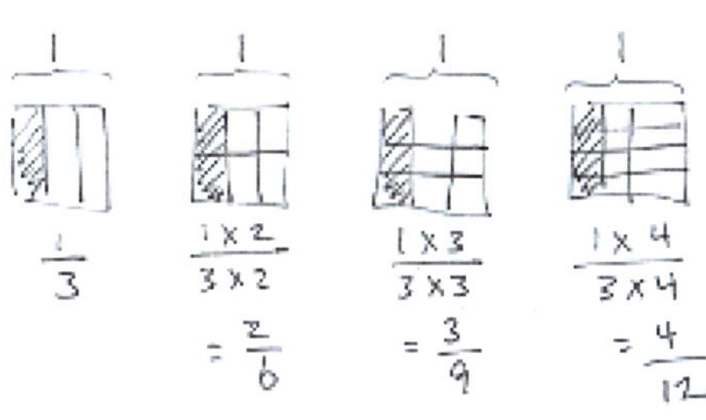

Problem 3: Make fractions equal to $\frac{2}{3}$.

The next complexity is working with a non-unit fraction.

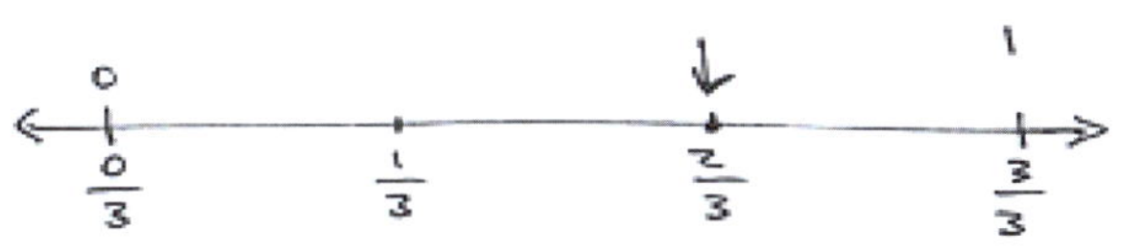

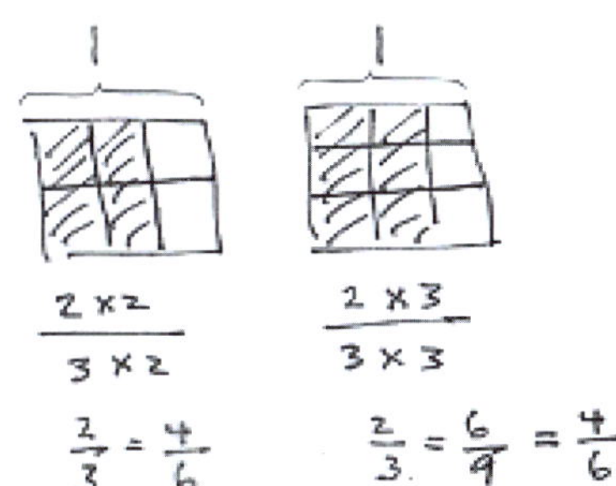

Problem 4: Make fractions equal to $\frac{5}{4}$.

Note: The final complexity prior to working independently is to model a fraction greater than 1. The same exact process is used. Rectangles are used in the example just to break rigidity. This is not unique to squares!

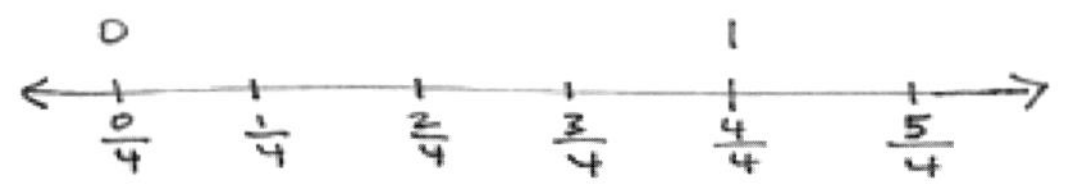

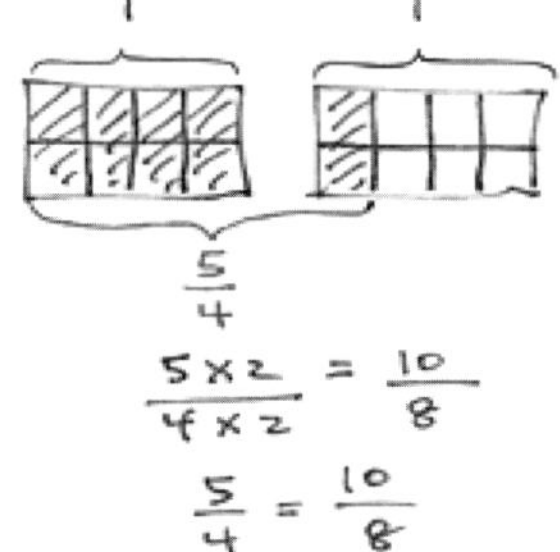

Problem Set (10 minutes)

Students should do their personal best to complete the Problem Set within the allotted 10 minutes. Some problems do not specify a method for solving. This is an intentional reduction of scaffolding that invokes MP.5, Use Appropriate Tools Strategically. Students should solve these problems using the RDW approach used for Application Problems.

For some classes, it may be appropriate to modify the assignment by specifying which problems students should work on first. With this option, let the purposeful sequencing of the Problem Set guide your selections so that problems continue to be scaffolded. Balance word problems with other problem types to ensure a range of practice. Consider assigning incomplete problems for homework or at another time during the day.

Name Jay Date

1. Use the folded paper strip to mark points 0 and 1 above the number line, and $\frac{0}{2}$, $\frac{1}{2}$, and $\frac{2}{2}$ below it.

Draw one vertical line down the middle of each rectangle, creating two parts. Shade the left half of each. Partition with horizontal lines to show the equivalent fractions $\frac{2}{4}$, $\frac{3}{6}$, $\frac{4}{8}$, and $\frac{5}{10}$. Use multiplication to show the change in the units.

$\frac{1}{2} = \frac{1 \times 2}{2 \times 2} = \frac{2}{4}$ $\frac{1}{2} = \frac{1 \times 3}{2 \times 3} = \frac{3}{6}$ $\frac{1}{2} = \frac{1 \times 4}{2 \times 4} = \frac{4}{8}$ $\frac{1}{2} = \frac{1 \times 5}{2 \times 5} = \frac{5}{10}$

2. Use the folded paper strip to mark points 0 and 1 above the number line, and $\frac{0}{3}$, $\frac{1}{3}$, $\frac{2}{3}$, and $\frac{3}{3}$ below it. Follow the same pattern as Problem 1, but with thirds.

$\frac{1}{3} = \frac{1 \times 2}{3 \times 2} = \frac{2}{6}$ $\frac{1}{3} = \frac{1 \times 3}{3 \times 3} = \frac{3}{9}$ $\frac{1}{3} = \frac{1 \times 4}{3 \times 4} = \frac{4}{12}$ $\frac{1}{3} = \frac{1 \times 5}{3 \times 5} = \frac{5}{15}$

Student Debrief (10 minutes)

Lesson Objective: Make equivalent fractions with the number line, the area model, and numbers.

The Student Debrief is intended to invite reflection and active processing of the total lesson experience.

Invite students to review their solutions for the Problem Set. They should check work by comparing answers with a partner before going over answers as a class. Look for misconceptions or misunderstandings that can be addressed in the Debrief. Guide students in a conversation to debrief the Problem Set and process the lesson.

T: Looking at your Problem Set, which fractions are equal to $\frac{1}{3}$?

S: $\frac{2}{6}, \frac{3}{9}, \frac{4}{12}, \frac{5}{15}$.

T: Continue the pattern beyond those on the Problem Set with your partner.

T: (After a moment.) Continue the pattern chorally.

S: $\frac{6}{18}, \frac{7}{21}, \frac{8}{24}, \frac{9}{27}, \frac{10}{30}$.

T: Is $\frac{100}{300}$ equal to $\frac{1}{3}$?

S: Yes.

T: How can we know if a fraction is equal to 1 third without drawing?

S: When you multiply the numerator by 3, you get the denominator. → When you divide the denominator by 3, you get the numerator. → The total number of equal pieces is 3 times the number of selected equal pieces.

T: In the next minute, write as many other fractions as you can that are equal to 1 third on your personal white board.

T: What do we know about all these fractions when we look at the number line?

S: They are the exact same point.

T: So, there are an infinite number of fractions equivalent to $\frac{1}{3}$?

S: Yes!

T: The fraction $\frac{1}{3}$ is one number, just like the number two or three. It is not two numbers, just this one point on the number line.

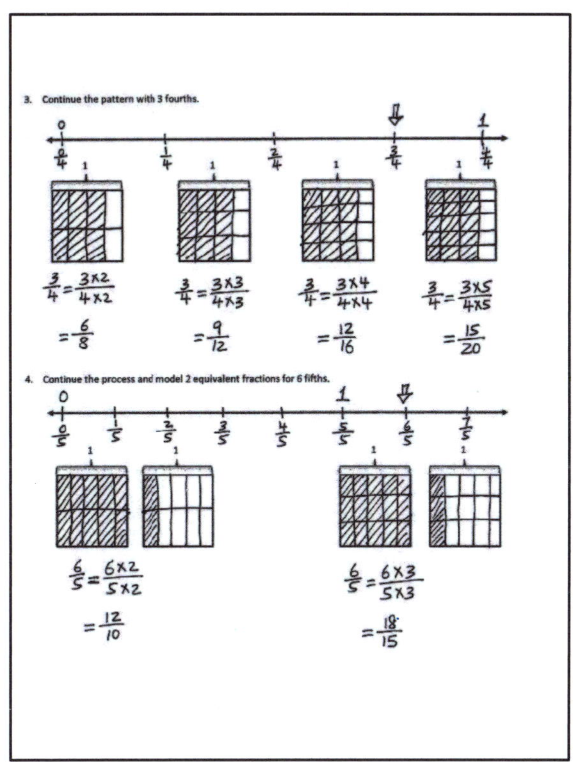
3. Continue the pattern with 3 fourths.

$\frac{3}{4} = \frac{3 \times 2}{4 \times 2} = \frac{6}{8}$ $\frac{3}{4} = \frac{3 \times 3}{4 \times 3} = \frac{9}{12}$ $\frac{3}{4} = \frac{3 \times 4}{4 \times 4} = \frac{12}{16}$ $\frac{3}{4} = \frac{3 \times 5}{4 \times 5} = \frac{15}{20}$

4. Continue the process and model 2 equivalent fractions for 6 fifths.

$\frac{6}{5} = \frac{6 \times 2}{5 \times 2} = \frac{12}{10}$ $\frac{6}{5} = \frac{6 \times 3}{5 \times 3} = \frac{18}{15}$

Quickly repeat the process of generating equivalent fractions for 3 fourths and 6 fifths.

MP.7

T: Discuss with your partner what is happening to the pieces—the units—when the numerator and denominator are getting larger.

S: The parts are getting smaller. → The equal pieces are being replaced by smaller equal pieces, but the area of the fraction is staying the same. → The units are being partitioned into smaller equal units. The value of the fraction is exactly the same.

NOTES ON MULTIPLE MEANS OF REPRESENTATION:

These number lines and squares are approximate sketches rather than precise drawings. Avoid rulers and graph paper so that students become accustomed to realizing that these images are not intended to be perfect but symbolic. The lines represent straight lines. The units are cut into equal parts. The Problem Set has pre-drawn squares in the interest of time and to expedite the movement to the abstract number. It is highly preferable to start with hand-drawn squares, so students do not receive the mistaken impression that drawings have to be perfect. A mental schema is developing, not an attachment to the drawing.

T: What would that look like were we to see it on the number line?

S: The length would be divided into smaller and smaller parts.

T: Discuss with your partner what the new smaller unit will be when I divide each of the lengths of 1 fourth into 3 smaller parts of equal length (use the fourths number line from earlier in the lesson). Compare it to the corresponding picture on your Problem Set.

MP.7

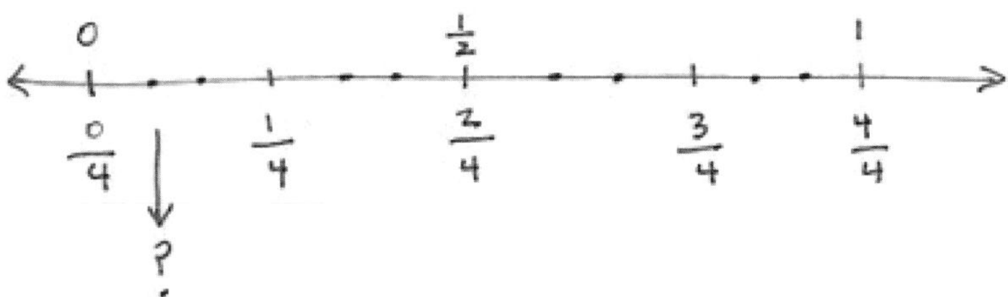

T: On your Problem Set, divide the lengths of 1 third into 5 equal smaller units. Think about what is happening to the units, length, and name of the fraction. Close the lesson by discussing connections between the number line, the area model, and the equivalent fractions with your partner.

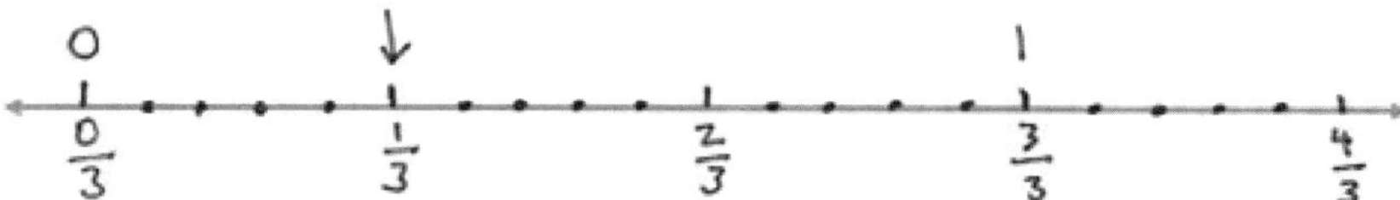

Exit Ticket (3 minutes)

After the Student Debrief, instruct students to complete the Exit Ticket. A review of their work will help with assessing students' understanding of the concepts that were presented in today's lesson and planning more effectively for future lessons. The questions may be read aloud to the students.

A

Number Correct: _______

Write the Missing Factor

1.	$10 = 5 \times$ ___		23.	$28 = 7 \times$ ___	
2.	$10 = 2 \times$ ___		24.	$28 = 2 \times 2 \times$ ___	
3.	$8 = 4 \times$ ___		25.	$28 = 2 \times$ ___ $\times 2$	
4.	$9 = 3 \times$ ___		26.	$28 =$ ___ $\times 2 \times 2$	
5.	$6 = 2 \times$ ___		27.	$36 = 3 \times 3 \times$ ___	
6.	$6 = 3 \times$ ___		28.	$9 \times 4 = 3 \times 3 \times$ ___	
7.	$12 = 6 \times$ ___		29.	$9 \times 4 = 6 \times$ ___	
8.	$12 = 3 \times$ ___		30.	$9 \times 4 = 3 \times 2 \times$ ___	
9.	$12 = 4 \times$ ___		31.	$8 \times 6 = 4 \times$ ___ $\times 2$	
10.	$12 = 2 \times 2 \times$ ___		32.	$9 \times 9 = 3 \times$ ___ $\times 3$	
11.	$12 = 3 \times 2 \times$ ___		33.	$8 \times 8 =$ ___ $\times 8$	
12.	$20 = 5 \times 2 \times$ ___		34.	$7 \times 7 =$ ___ $\times 7$	
13.	$20 = 5 \times 2 \times$ ___		35.	$8 \times 3 =$ ___ $\times 6$	
14.	$16 = 8 \times$ ___		36.	$16 \times 2 =$ ___ $\times 4$	
15.	$16 = 4 \times 2 \times$ ___		37.	$2 \times 18 =$ ___ $\times 9$	
16.	$24 = 8 \times$ ___		38.	$28 \times 2 =$ ___ $\times 8$	
17.	$24 = 4 \times 2 \times$ ___		39.	$24 \times 3 =$ ___ $\times 9$	
18.	$24 = 4 \times$ ___ $\times 2$		40.	$6 \times 8 =$ ___ $\times 12$	
19.	$24 = 3 \times 2 \times$ ___		41.	$27 \times 3 =$ ___ $\times 9$	
20.	$24 = 3 \times$ ___ $\times 2$		42.	$12 \times 6 =$ ___ $\times 8$	
21.	$6 \times 4 = 8 \times$ ___		43.	$54 \times 2 =$ ___ $\times 12$	
22.	$6 \times 4 = 4 \times 2 \times$ ___		44.	$9 \times 13 =$ ___ $\times 39$	

B

Number Correct: _______

Improvement: _______

Write the Missing Factor

1.	6 = 2 × ___			23.	28 = 4 × ___	
2.	6 = 3 × ___			24.	28 = 2 × 2 × ___	
3.	9 = 3 × ___			25.	28 = 2 × ___ × 2	
4.	8 = 4 × ___			26.	28 = ___ × 2 × 2	
5.	10 = 5 × ___			27.	36 = 2 × 2 × ___	
6.	10 = 2 × ___			28.	9 × 4 = 2 × 2 × ___	
7.	20 = 10 × ___			29.	9 × 4 = 6 × ___	
8.	20 = 5 × 2 × ___			30.	9 × 4 = 2 × 3 × ___	
9.	12 = 6 × ___			31.	8 × 6 = 4 × ___ × 2	
10.	12 = 3 × ___			32.	8 × 8 = 4 × ___ × 2	
11.	12 = 4 × ___			33.	9 × 9 = ___ × 9	
12.	12 = 2 × 2 × ___			34.	6 × 6 = ___ × 6	
13.	12 = 3 × 2 × ___			35.	6 × 4 = ___ × 8	
14.	24 = 8 × ___			36.	16 × 2 = ___ × 8	
15.	24 = 4 × 2 × ___			37.	2 × 18 = ___ × 4	
16.	24 = 4 × ___ × 2			38.	28 × 2 = ___ × 7	
17.	24 = 3 × 2 × ___			39.	24 × 3 = ___ × 8	
18.	24 = 3 × ___ × 2			40.	8 × 6 = ___ × 4	
19.	16 = 8 × ___			41.	12 × 6 = ___ × 9	
20.	16 = 4 × 2 × ___			42.	27 × 3 = ___ × 9	
21.	8 × 2 = 4 × ___			43.	54 × 2 = ___ × 9	
22.	8 × 2 = 2 × 2 × ___			44.	8 × 13 = ___ × 26	

Name ______________________________ Date ________________

1. Use the folded paper strip to mark points 0 and 1 above the number line and $\frac{0}{2}$, $\frac{1}{2}$, and $\frac{2}{2}$ below it.

Draw one vertical line down the middle of each rectangle, creating two parts. Shade the left half of each. Partition with horizontal lines to show the equivalent fractions $\frac{2}{4}$, $\frac{3}{6}$, $\frac{4}{8}$, and $\frac{5}{10}$. Use multiplication to show the change in the units.

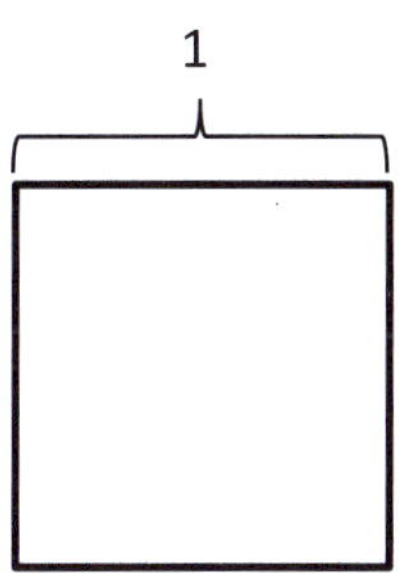

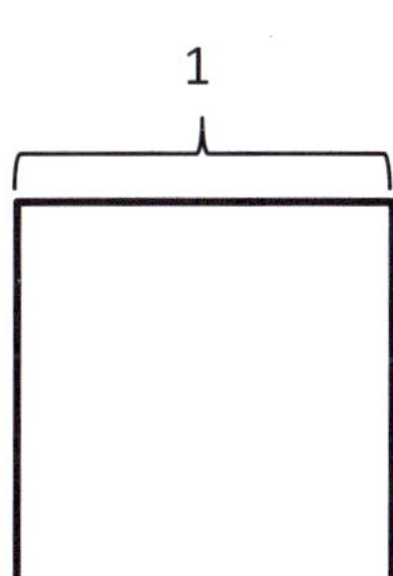

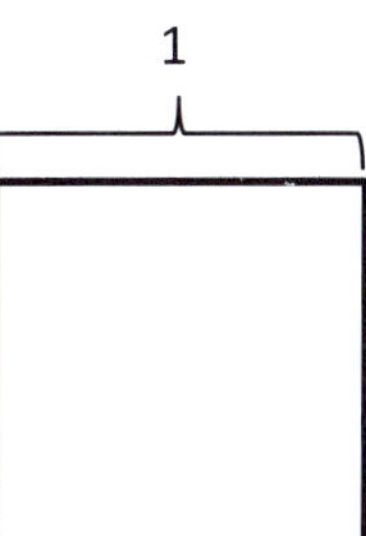

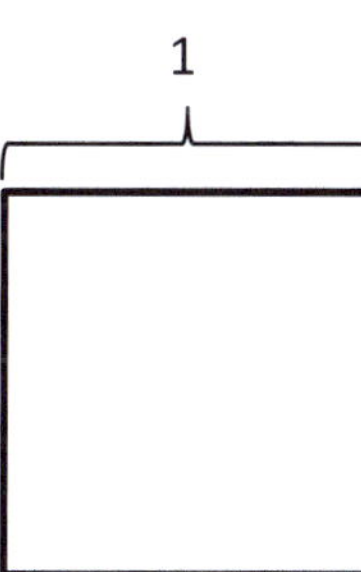

$$\frac{1}{2} = \frac{1 \times 2}{2 \times 2} = \frac{2}{4}$$

2. Use the folded paper strip to mark points 0 and 1 above the number line and $\frac{0}{3}$, $\frac{1}{3}$, $\frac{2}{3}$, and $\frac{3}{3}$ below it. Follow the same pattern as Problem 1 but with thirds.

1

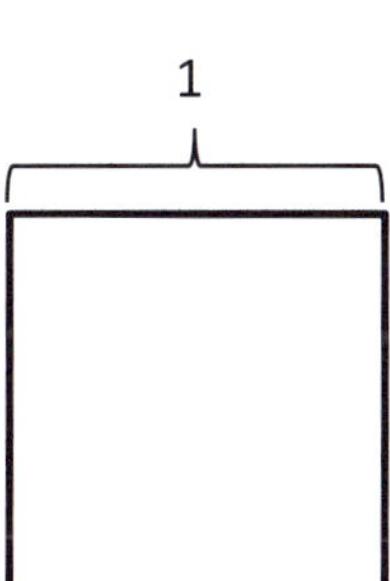

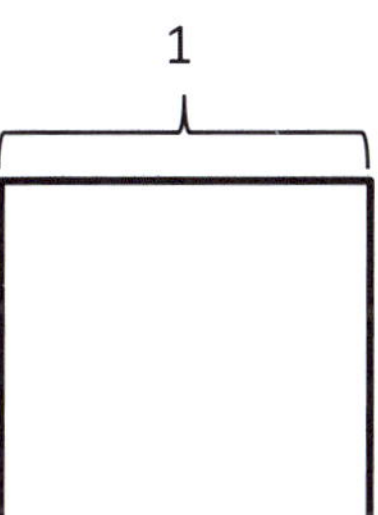

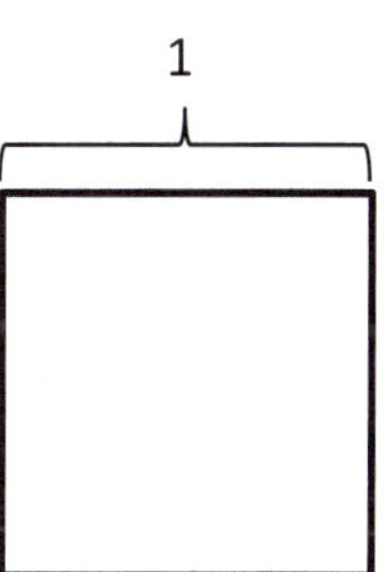

3. Continue the pattern with 3 fourths.

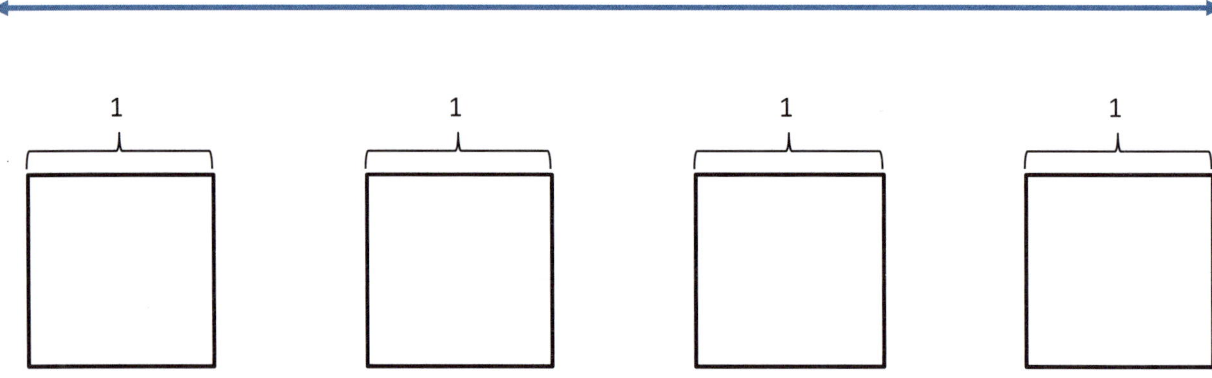

4. Continue the process, and model 2 equivalent fractions for 6 fifths.

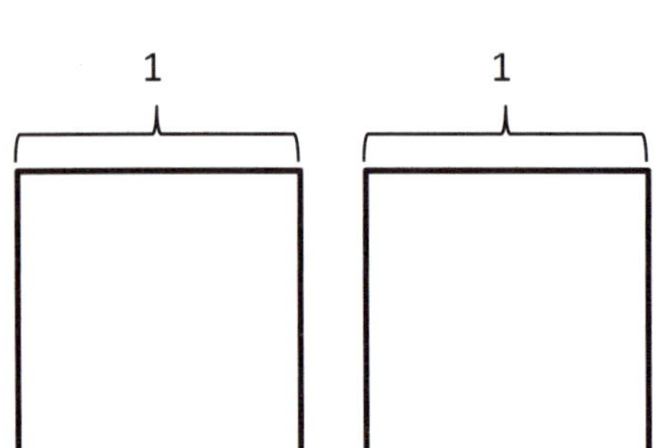

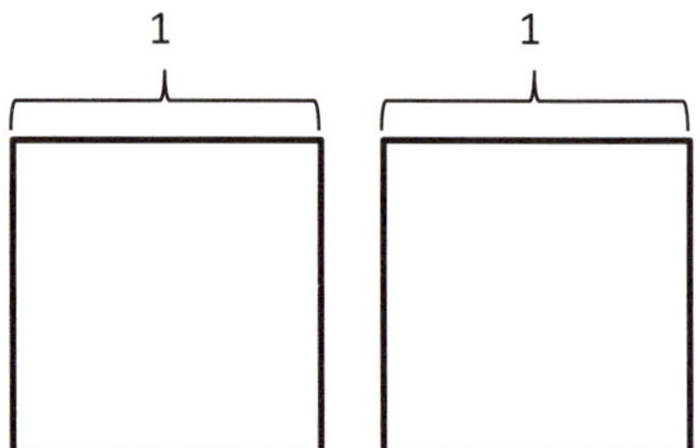

Name ______________________________ Date ______________

Estimate to mark points 0 and 1 above the number line, and $\frac{0}{6}$, $\frac{1}{6}$, $\frac{2}{6}$, $\frac{3}{6}$, $\frac{4}{6}$, $\frac{5}{6}$, and $\frac{6}{6}$ below it. Use the squares below to represent fractions equivalent to 1 sixth using both arrays and equations.

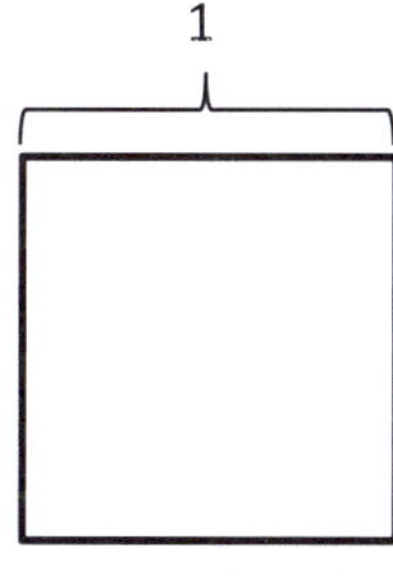

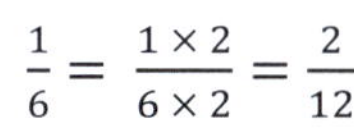

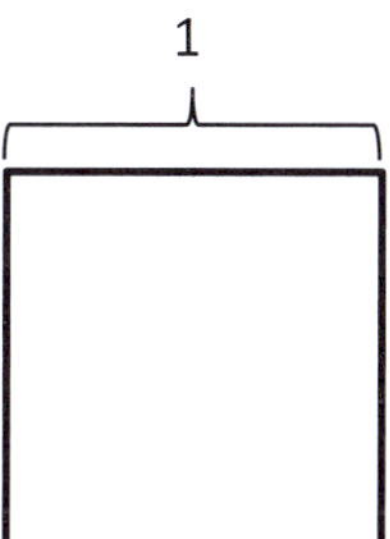

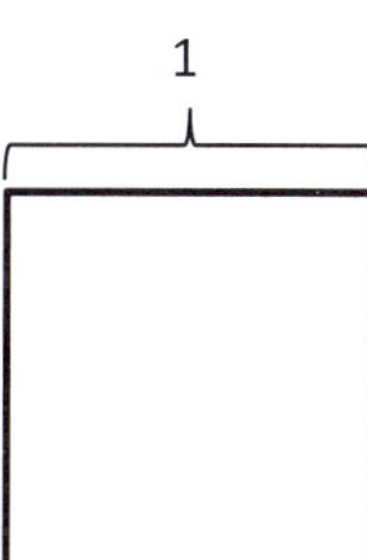

Name ______________________________ Date ______________

1. Use the folded paper strip to mark points 0 and 1 above the number line and $\frac{0}{3}$, $\frac{1}{3}$, $\frac{2}{3}$, and $\frac{3}{3}$ below it.

Draw two vertical lines to break each rectangle into thirds. Shade the left third of each. Partition with horizontal lines to show equivalent fractions. Use multiplication to show the change in the units.

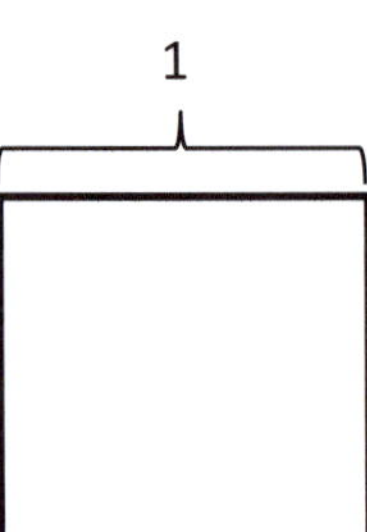

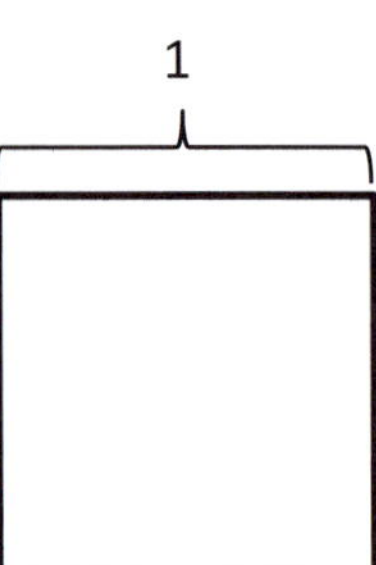

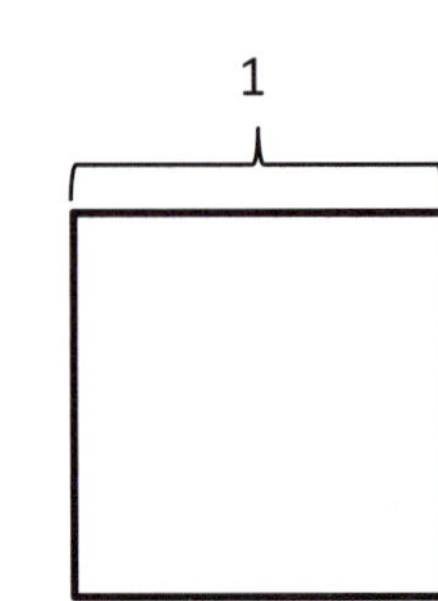

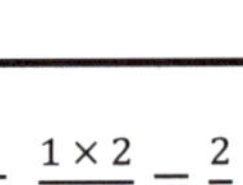

2. Use the folded paper strip to mark points 0 and 1 above the number line and $\frac{0}{4}, \frac{1}{4}, \frac{2}{4}, \frac{3}{4}$, and $\frac{4}{4}$ below it. Follow the same pattern as Problem 1 but with fourths.

1

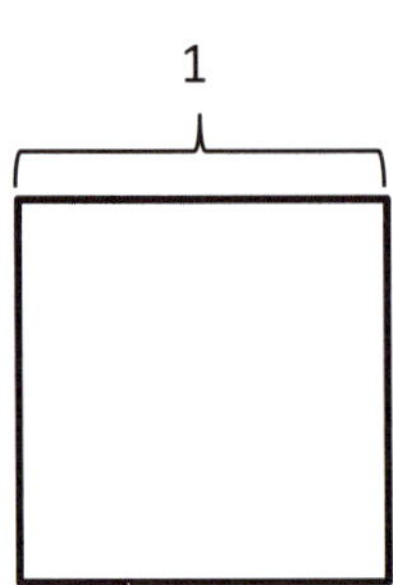

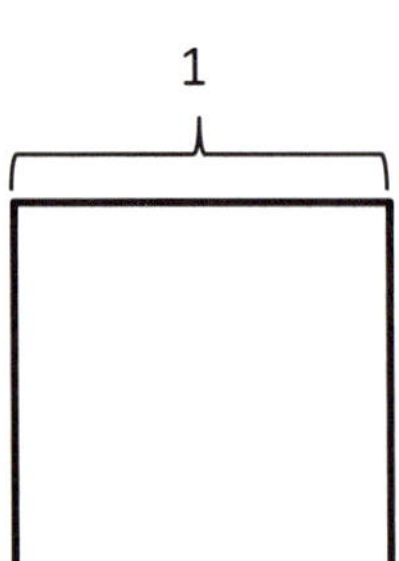

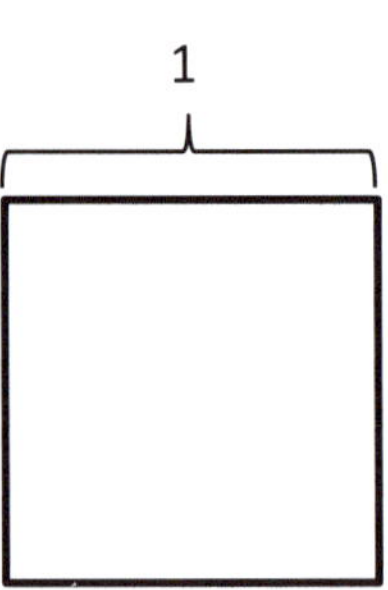

3. Continue the pattern with 4 fifths.

1 1 1 1

4. Continue the process, and model 2 equivalent fractions for 9 eighths. Estimate to mark the points on the number line.

1 1

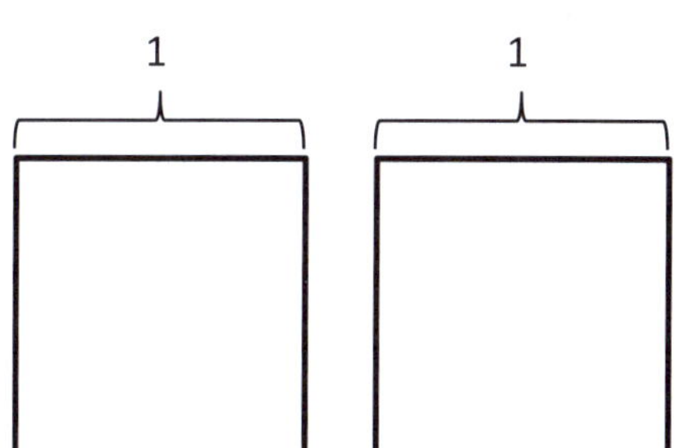

Lesson 2

Objective: Make equivalent fractions with sums of fractions with like denominators.

Suggested Lesson Structure

Fluency Practice	(12 minutes)
Application Problem	(8 minutes)
Concept Development	(30 minutes)
Student Debrief	(10 minutes)
Total Time	**(60 minutes)**

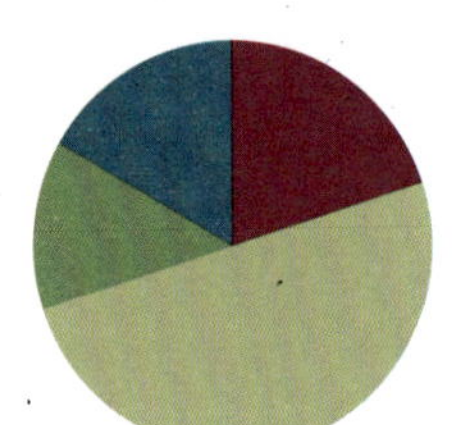

NOTES ON MULTIPLE MEANS OF ENGAGEMENT:

Equivalent Fractions is intentionally placed before the Sprint because it reviews the Sprint skill. Adjust the time for the Sprint as necessary. If students struggle to complete Sprint A, then consider doing another minute or two of Equivalent Fractions before moving them on to Sprint B.

Fluency Practice (12 minutes)

- Equivalent Fractions **5.NF.1** (4 minutes)
- Sprint: Find the Missing Numerator or Denominator **4.NF.1** (8 minutes)

Equivalent Fractions (4 minutes)

Note: This fluency activity reviews equivalent fractions.

T: (Write $\frac{1}{2}$.) Say the fraction.

S: One half.

T: (Write $\frac{1}{2} = \frac{}{4}$.) One half is how many fourths?

S: Two fourths.

Continue with the following possible sequence:

$\frac{1}{2} = \frac{}{6}$, $\frac{1}{3} = \frac{}{6}$, $\frac{2}{3} = \frac{}{6}$, $\frac{2}{3} = \frac{}{12}$, $\frac{3}{4} = \frac{}{16}$, and $\frac{3}{5} = \frac{}{25}$.

T: (Write $\frac{1}{2}$.) Say the fraction.

S: One half.

T: (Write $\frac{1}{2} = \frac{2}{}$.) One half or one part of two is the same as two parts of what unit?

S: Fourths.

Continue with the following possible sequence:

$\frac{1}{2} = \frac{2}{}$, $\frac{1}{5} = \frac{2}{}$, $\frac{2}{5} = \frac{8}{}$, $\frac{3}{4} = \frac{9}{}$, and $\frac{4}{5} = \frac{16}{}$.

NOTES ON MULTIPLE MEANS OF REPRESENTATION:

Adjusting number words and correctly pronouncing them as fractions (fifths, sixths, etc.) may be challenging. If there are many English language learners in class, consider quickly counting together to practice enunciating word endings: hal**ves**, thir**ds**, four**ths**, fif**ths**, six**ths**, etc.

Sprint: Find the Missing Numerator or Denominator (8 minutes)

Materials: (S) Find the Missing Numerator or Denominator Sprint

Note: Students generate common equivalent fractions mentally and with automaticity (i.e., without performing the indicated multiplication).

Application Problem (8 minutes)

Mr. Hopkins has a 1-meter wire he is using to make clocks. Each fourth meter is marked off and divided into 5 smaller equal lengths. If Mr. Hopkins bends the wire at $\frac{3}{4}$ meter, what fraction of the smaller marks is that?

S: (Solve the problem, possibly using the RDW process independently or in partners.)

T: Let's look at two of your solutions and compare them.

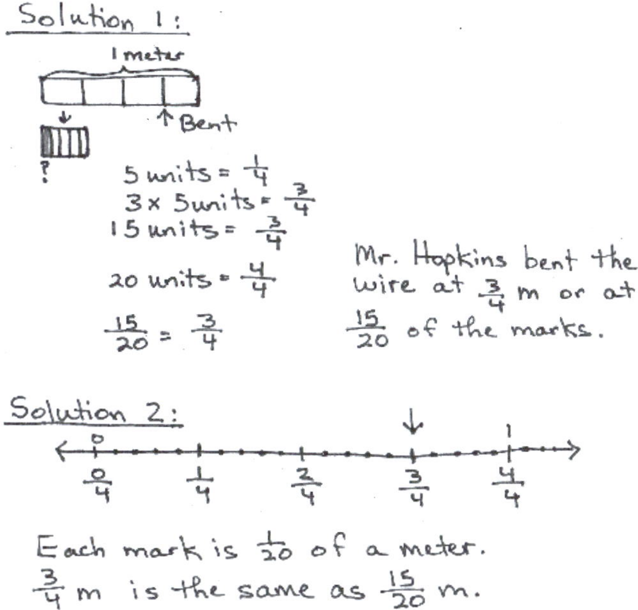

NOTES ON SOLVING APPLICATION PROBLEMS:

Since Grade 1, students have used the Read, Draw, Write (RDW) approach to solve Application Problems. The method is as follows:

1. Read the problem.
2. Draw to represent the problem.
3. Write one or more equations that either help solve the problem or show how the problem was solved.
4. Write a statement that answers the question.

Embedded within Draw are important reflective questions:

- What do I see?
- Can I draw something?
- What conclusions can I reach from my drawing?

T: When you look at these two solutions side by side, what do you see? (Consider using the following set of questions to help students compare the solutions as a whole class or to encourage inter-partner communication while circulating as they compare.)

- What did each of these students draw?
- What conclusions can you make from their drawings?
- How did they record their solutions numerically?
- How does the tape diagram relate to the number line?
- What does the tape diagram/number line clarify?

- What does the equation clarify?
- How could the statement with the number line be rephrased to answer the question?

Note: This two-step Application Problem offers a problem-solving context for students to review making equivalent fractions with the number line or the area model as taught in Lesson 1.

Concept Development (30 minutes)

Materials: (S) Blank paper

Problem 1: $\frac{1}{3}+\frac{1}{3}=\frac{2}{3}$.

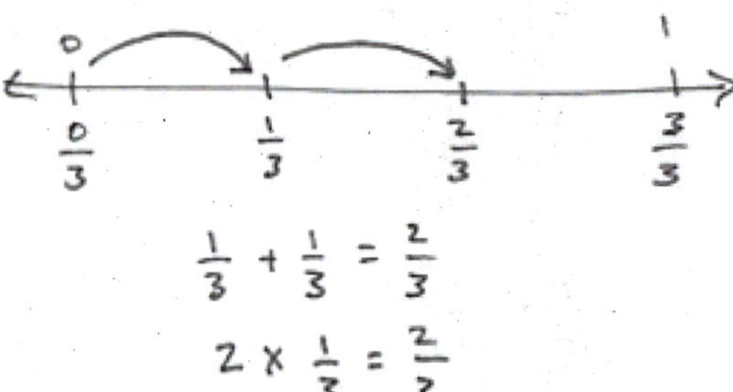

1 third + 1 third = 2 thirds.

T: Draw a number line. Mark the endpoints as 0 and 1. Between zero and one, estimate to make three units of equal length and label them as thirds.

S: (Work.)

T: On your number line, show 1 third plus 1 third with arrows designating lengths. (Demonstrate, and then pause as students work.)

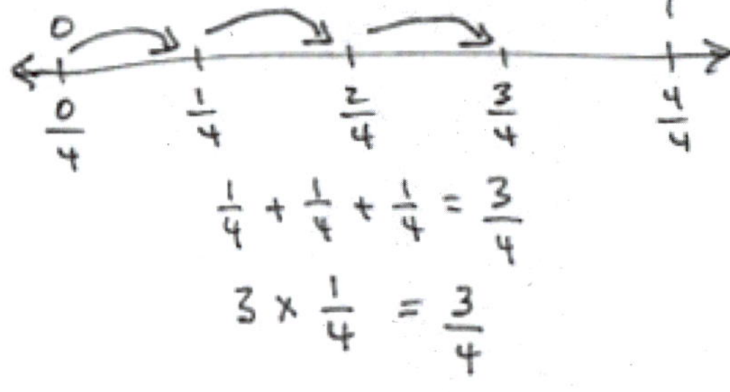

T: The answer is ...?

S: 2 thirds.

T: Talk to your partner. Express this as an addition sentence and a multiplication equation.

S: $\frac{1}{3}+\frac{1}{3}=\frac{2}{3}$. → $2\times\frac{1}{3}=\frac{2}{3}$.

T: Following the same pattern of adding unit fractions by joining lengths, show 3 fourths on a number line.

Problem 2: $\frac{3}{8}+\frac{3}{8}+\frac{1}{8}=\frac{7}{8}$.

3 eighths + 3 eighths + 1 eighth = 7 eighths.

T: Draw a number line. Again, mark the endpoints as 0 and 1. Between zero and one, estimate to make eight units of equal length. This time, only label what is necessary to show 3 eighths.

S: (Work.)

T: Represent 3 eighths + 3 eighths + 1 eighth on your number line. (Pause.) What's the answer?

S: 7 eighths.

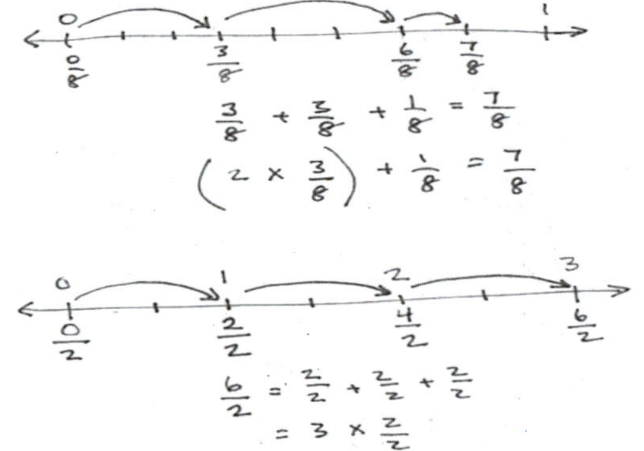

T: Talk to your partner. Express this as an addition equation and a multiplication equation.

S: $\frac{3}{8}+\frac{3}{8}+\frac{1}{8}=\frac{7}{8}$. → $\left(2\times\frac{3}{8}\right)+\frac{1}{8}=\frac{7}{8}$.

Problem 3: $\frac{6}{2}=\frac{2}{2}+\frac{2}{2}+\frac{2}{2}=\left(3\times\frac{2}{2}\right)=3.$

6 halves = 3 × 2 halves = 3 ones = 3.

T: Draw a number line. Below the number line, mark the endpoints as 0 halves and 6 halves. Estimate to make 6 parts of equal length. This time, only label 2 halves.

S: (Work.)

T: Record the whole number equivalents above the line. (Record 1, 2, and 3 wholes.) Represent 3 × 2 halves on your number line.

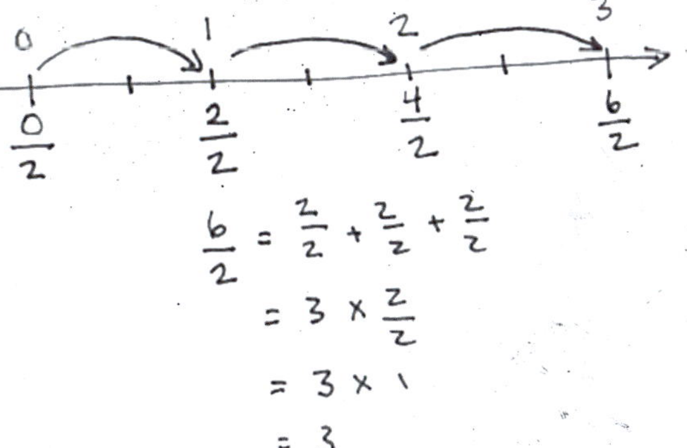

S: (Draw 3 arrows, starting with $\frac{0}{2}$, $\frac{2}{2}$, $\frac{4}{2}$, and stop at $\frac{6}{2}$.)

T: What's the answer?

S: 6 halves or 3.

T: 3. What is the unit?

S: 3 ones.

T: Talk to your partner. Express this as an addition equation and a multiplication equation.

S: $\frac{2}{2}+\frac{2}{2}+\frac{2}{2}=\frac{6}{2}=3.$ → $\frac{2}{2}+\frac{2}{2}+\frac{2}{2}=1+1+1=3.$ → $3\times\frac{2}{2}=\frac{6}{2}=3.$ → $3\times\frac{2}{2}=3\times1=3.$

Problem 4: $\frac{8}{5}=\frac{5}{5}+\frac{3}{5}=1\frac{3}{5}.$

8 fifths = 5 fifths + 3 fifths = 1 and 3 fifths.

T: Draw a number line. Below the number line, mark the endpoints as 0 fifths and 10 fifths. Estimate and give a value to the halfway point.

T: What is the value of the halfway point?

S: 5 fifths.

T: Make 10 parts of equal length from 0 fifths to 10 fifths.

T: Record the whole number equivalents above the line. (Students work.)

T: Label 8 fifths on your number line. (Work.)

T: Show 8 fifths as the sum of 5 fifths and 3 fifths on your number line.

S: (Work.)

T: Talk to your partner. Express this as an addition equation in two ways—as the sum of fifths as well as the sum of a whole number and fifths.

S: (Work.)

T: What is another way of expressing 1 plus 3 fifths?

S: 1 and 3 fifths. → $\frac{5}{5} + \frac{3}{5} = \frac{8}{5} = 1\frac{3}{5}$.

T: 8 fifths is between what 2 whole numbers?

S: 1 and 2.

Problem 5: $\frac{7}{3} = \frac{6}{3} + \frac{1}{3} = \left(2 \times \frac{3}{3}\right) + \frac{1}{3} = 2 + \frac{1}{3} = 2\frac{1}{3}$.

7 thirds = 6 thirds + 1 third = 2 and 1 third.

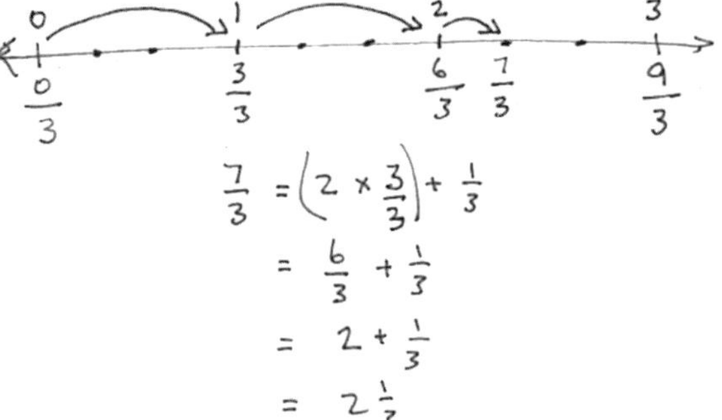

T: Draw a number line. Mark the endpoints as 0 thirds and 9 thirds below the number line. Divide the whole length into three equal smaller lengths and mark their values using thirds. Work with a partner.

S: (Work.)

T: What are the values of those points?

S: 3 thirds and 6 thirds.

T: Mark the whole number equivalents above the line.

S: (Work.)

T: Divide each of those whole number lengths into three smaller lengths. Mark the number 7 thirds.

S: (Work.)

T: Show 7 thirds as two units of 3 thirds and one more third on your number line and in an equation. Work together if you become stuck.

S: (Work and discuss.)

T: 7 thirds is between what two whole numbers?

S: 2 and 3.

Problem Set (10 minutes)

Students should do their personal best to complete the Problem Set within the allotted 10 minutes. For some classes, it may be appropriate to modify the assignment specifying which problems they work on first. Some problems do not specify a method for solving. Students should solve these problems using the RDW approach used for Application Problems.

Student Debrief (10 minutes)

Lesson Objective: Make equivalent fractions with sums of fractions with like denominators.

The Student Debrief is intended to invite reflection and active processing of the total lesson experience.

Invite students to review their solutions for the Problem Set. They should check work by comparing answers with a partner before going over answers as a class. Look for misconceptions or misunderstandings that can be addressed in the Debrief. Guide students in a conversation to debrief the Problem Set and process the lesson.

T: Come to the Debrief and bring your Problem Set. Compare your work to your neighbor's. On which problems do you have different answers? Discuss your differences. Both may be correct.

T: (After about 3 minutes.) What is a way to express $\frac{3}{7}$ as a sum?

S: 1 seventh + 1 seventh + 1 seventh.

T: Another way?

S: 2 sevenths + 1 seventh.

T: These are equivalent forms of 3 sevenths.

T: On your Problem Set, find and talk to your partner about different equivalent forms of your numbers.

S: 6 sevenths could be expressed as 3 sevenths + 3 sevenths or 3 times 2 sevenths. → 9 sevenths can be expressed as 1 + 2 sevenths. → 7 fourths can be expressed as 2 times 3 fourths + 1 fourth. → 1 and 3 fourths can be expressed as 7 fourths. → 32 sevenths can be expressed as 28 sevenths + 4 sevenths or 4 and 4 sevenths.

T: I'm hearing you express these numbers in many equivalent forms. Why do you think I chose to use the tool of the number line in this lesson? Discuss this with your partner. If you were the teacher of this lesson, why might you use the number line?

S: (Discuss.)

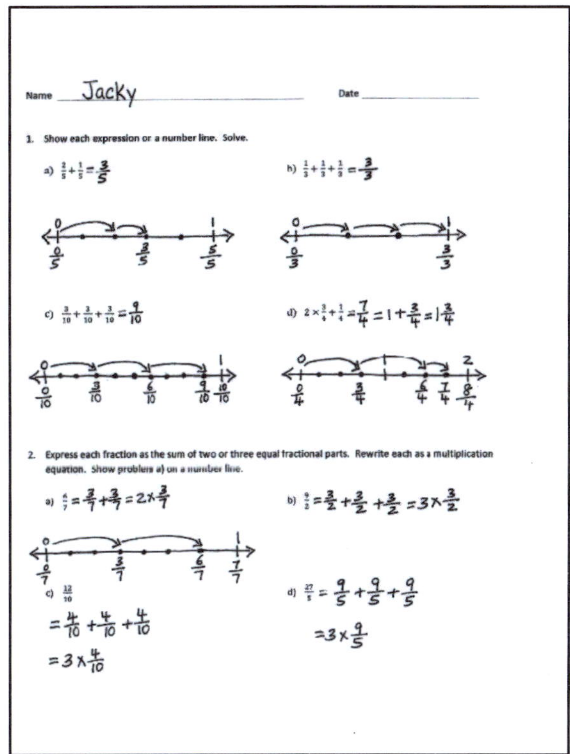

Name Jacky Date

1. Show each expression or a number line. Solve.

a) $\frac{2}{5}+\frac{1}{5}=\frac{3}{5}$

h) $\frac{1}{3}+\frac{1}{3}+\frac{1}{3}=\frac{3}{3}$

c) $\frac{3}{10}+\frac{3}{10}+\frac{3}{10}=\frac{9}{10}$

d) $2\times\frac{3}{4}+\frac{1}{4}=\frac{7}{4}=1+\frac{3}{4}=1\frac{3}{4}$

2. Express each fraction as the sum of two or three equal fractional parts. Rewrite each as a multiplication equation. Show problem a) on a number line.

a) $\frac{6}{7}=\frac{3}{7}+\frac{3}{7}=2\times\frac{3}{7}$

b) $\frac{9}{2}=\frac{3}{2}+\frac{3}{2}+\frac{3}{2}=3\times\frac{3}{2}$

c) $\frac{12}{10}$ $=\frac{4}{10}+\frac{4}{10}+\frac{4}{10}$ $=3\times\frac{4}{10}$

d) $\frac{27}{5}=\frac{9}{5}+\frac{9}{5}+\frac{9}{5}$ $=3\times\frac{9}{5}$

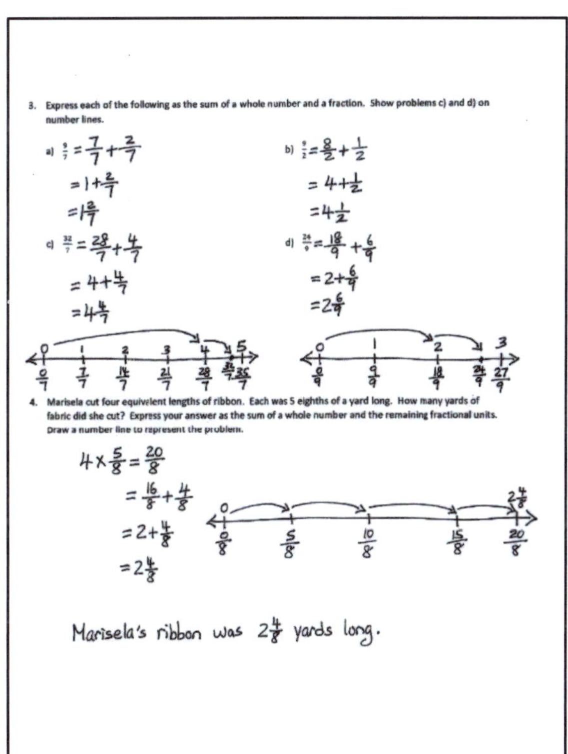

3. Express each of the following as the sum of a whole number and a fraction. Show problems c) and d) on number lines.

a) $\frac{9}{7}=\frac{7}{7}+\frac{2}{7}$ $=1+\frac{2}{7}$ $=1\frac{2}{7}$

b) $\frac{9}{2}=\frac{8}{2}+\frac{1}{2}$ $=4+\frac{1}{2}$ $=4\frac{1}{2}$

c) $\frac{32}{7}=\frac{28}{7}+\frac{4}{7}$ $=4+\frac{4}{7}$ $=4\frac{4}{7}$

d) $\frac{24}{9}=\frac{18}{9}+\frac{6}{9}$ $=2+\frac{6}{9}$ $=2\frac{6}{9}$

4. Marisela cut four equivalent lengths of ribbon. Each was 5 eighths of a yard long. How many yards of fabric did she cut? Express your answer as the sum of a whole number and the remaining fractional units. Draw a number line to represent the problem.

$4\times\frac{5}{8}=\frac{20}{8}$ $=\frac{16}{8}+\frac{4}{8}$ $=2+\frac{4}{8}$ $=2\frac{4}{8}$

Marisela's ribbon was $2\frac{4}{8}$ yards long.

T: When we were studying decimal place value, we saw that 9 tenths + 3 tenths is equal to 12 tenths, 1 + 2 tenths, or 1 and 2 tenths.

T: Once more, please review the solution and number line you made for Problem 4 about Marisela's ribbon. Discuss the equivalence of 20 eighths and 2 and 4 eighths as it relates to the number line.

S: (Discuss.)

T: Discuss the relationship of the equivalence of these sums.

9 tenths + 3 tenths = 12 tenths = 1 + 2 tenths = $1\frac{2}{10}$.

9 elevenths + 3 elevenths = 12 elevenths = 1 + 1 eleventh = $1\frac{1}{11}$.

S: (Discuss.)

T: Yes, our place value system is another example of equivalence.

Exit Ticket (3 minutes)

After the Student Debrief, instruct students to complete the Exit Ticket. A review of their work will help with assessing students' understanding of the concepts that were presented in today's lesson and planning more effectively for future lessons. The questions may be read aloud to the students.

A

Number Correct: _______

Find the Missing Numerator or Denominator

1.	$\frac{1}{2} = \frac{}{4}$	
2.	$\frac{1}{5} = \frac{2}{}$	
3.	$\frac{2}{5} = \frac{}{10}$	
4.	$\frac{3}{5} = \frac{}{10}$	
5.	$\frac{4}{5} = \frac{}{10}$	
6.	$\frac{1}{3} = \frac{2}{}$	
7.	$\frac{2}{3} = \frac{}{6}$	
8.	$\frac{1}{3} = \frac{3}{}$	
9.	$\frac{2}{3} = \frac{}{9}$	
10.	$\frac{1}{4} = \frac{}{8}$	
11.	$\frac{3}{4} = \frac{}{8}$	
12.	$\frac{1}{4} = \frac{3}{}$	
13.	$\frac{3}{4} = \frac{9}{}$	
14.	$\frac{2}{4} = \frac{}{2}$	
15.	$\frac{2}{6} = \frac{1}{}$	
16.	$\frac{2}{10} = \frac{1}{}$	
17.	$\frac{4}{10} = \frac{}{5}$	
18.	$\frac{8}{10} = \frac{}{5}$	
19.	$\frac{3}{9} = \frac{}{3}$	
20.	$\frac{6}{9} = \frac{}{3}$	
21.	$\frac{3}{12} = \frac{1}{}$	
22.	$\frac{9}{12} = \frac{}{4}$	

23.	$\frac{1}{3} = \frac{}{12}$	
24.	$\frac{2}{3} = \frac{}{12}$	
25.	$\frac{8}{12} = \frac{}{3}$	
26.	$\frac{12}{16} = \frac{3}{}$	
27.	$\frac{3}{5} = \frac{}{25}$	
28.	$\frac{4}{5} = \frac{28}{}$	
29.	$\frac{18}{24} = \frac{3}{}$	
30.	$\frac{24}{30} = \frac{}{5}$	
31.	$\frac{5}{6} = \frac{35}{}$	
32.	$\frac{56}{63} = \frac{}{9}$	
33.	$\frac{64}{72} = \frac{8}{}$	
34.	$\frac{5}{8} = \frac{}{64}$	
35.	$\frac{5}{6} = \frac{45}{}$	
36.	$\frac{45}{81} = \frac{}{9}$	
37.	$\frac{6}{7} = \frac{48}{}$	
38.	$\frac{36}{81} = \frac{}{9}$	
39.	$\frac{8}{56} = \frac{1}{}$	
40.	$\frac{35}{63} = \frac{5}{}$	
41.	$\frac{1}{6} = \frac{12}{}$	
42.	$\frac{3}{7} = \frac{36}{}$	
43.	$\frac{48}{60} = \frac{4}{}$	
44.	$\frac{72}{84} = \frac{}{7}$	

B

Number Correct: _______

Improvement: _______

Find the Missing Numerator or Denominator

#	Problem		#	Problem	
1.	$\frac{1}{5} = \frac{2}{\ }$		23.	$\frac{1}{3} = \frac{4}{\ }$	
2.	$\frac{2}{5} = \frac{\ }{10}$		24.	$\frac{2}{3} = \frac{8}{\ }$	
3.	$\frac{3}{5} = \frac{\ }{10}$		25.	$\frac{8}{12} = \frac{2}{\ }$	
4.	$\frac{4}{5} = \frac{\ }{10}$		26.	$\frac{12}{16} = \frac{\ }{4}$	
5.	$\frac{1}{3} = \frac{2}{\ }$		27.	$\frac{3}{5} = \frac{15}{\ }$	
6.	$\frac{1}{3} = \frac{\ }{6}$		28.	$\frac{4}{5} = \frac{\ }{35}$	
7.	$\frac{2}{3} = \frac{4}{\ }$		29.	$\frac{18}{24} = \frac{\ }{4}$	
8.	$\frac{1}{3} = \frac{\ }{9}$		30.	$\frac{24}{30} = \frac{4}{\ }$	
9.	$\frac{2}{3} = \frac{6}{\ }$		31.	$\frac{5}{6} = \frac{\ }{42}$	
10.	$\frac{1}{4} = \frac{2}{\ }$		32.	$\frac{56}{63} = \frac{8}{\ }$	
11.	$\frac{3}{4} = \frac{6}{\ }$		33.	$\frac{64}{72} = \frac{\ }{9}$	
12.	$\frac{1}{4} = \frac{\ }{12}$		34.	$\frac{5}{8} = \frac{40}{\ }$	
13.	$\frac{3}{4} = \frac{\ }{12}$		35.	$\frac{5}{6} = \frac{\ }{54}$	
14.	$\frac{2}{4} = \frac{1}{\ }$		36.	$\frac{45}{81} = \frac{5}{\ }$	
15.	$\frac{2}{6} = \frac{\ }{3}$		37.	$\frac{6}{7} = \frac{\ }{56}$	
16.	$\frac{2}{10} = \frac{\ }{5}$		38.	$\frac{36}{81} = \frac{4}{\ }$	
17.	$\frac{4}{10} = \frac{2}{\ }$		39.	$\frac{8}{56} = \frac{\ }{7}$	
18.	$\frac{8}{10} = \frac{4}{\ }$		40.	$\frac{35}{63} = \frac{\ }{9}$	
19.	$\frac{3}{9} = \frac{1}{\ }$		41.	$\frac{1}{6} = \frac{\ }{72}$	
20.	$\frac{6}{9} = \frac{2}{\ }$		42.	$\frac{3}{7} = \frac{\ }{84}$	
21.	$\frac{1}{4} = \frac{\ }{12}$		43.	$\frac{48}{60} = \frac{\ }{5}$	
22.	$\frac{9}{12} = \frac{3}{\ }$		44.	$\frac{72}{84} = \frac{6}{\ }$	

Name ______________________________ Date ______________

1. Show each expression on a number line. Solve.

a. $\frac{2}{5}+\frac{1}{5}$

b. $\frac{1}{3}+\frac{1}{3}+\frac{1}{3}$

c. $\frac{3}{10}+\frac{3}{10}+\frac{3}{10}$

d. $2\times\frac{3}{4}+\frac{1}{4}$

2. Express each fraction as the sum of two or three equal fractional parts. Rewrite each as a multiplication equation. Show Part (a) on a number line.

a. $\frac{6}{7}$

b. $\frac{9}{2}$

c. $\frac{12}{10}$

d. $\frac{27}{5}$

3. Express each of the following as the sum of a whole number and a fraction. Show Parts (c) and (d) on number lines.

   a. $\frac{9}{7}$

   b. $\frac{9}{2}$

   c. $\frac{32}{7}$

   d. $\frac{24}{9}$

4. Marisela cut four equivalent lengths of ribbon. Each was 5 eighths of a yard long. How many yards of ribbon did she cut? Express your answer as the sum of a whole number and the remaining fractional units. Draw a number line to represent the problem.

Name ______________________________ Date ______________

1. Show each expression on a number line. Solve.

   a. $\frac{5}{5}+\frac{2}{5}$

   b. $\frac{6}{3}+\frac{2}{3}$

2. Express each fraction as the sum of two or three equal fractional parts. Rewrite each as a multiplication equation. Show Part (b) on a number line.

   a. $\frac{6}{9}$

   b. $\frac{15}{4}$

Name ______________________________ Date ______________

1. Show each expression on a number line. Solve.

a. $\frac{4}{9}+\frac{1}{9}$

b. $\frac{1}{4}+\frac{1}{4}+\frac{1}{4}+\frac{1}{4}$

c. $\frac{2}{7}+\frac{2}{7}+\frac{2}{7}$

d. $2\times\frac{3}{5}+\frac{1}{5}$

2. Express each fraction as the sum of two or three equal fractional parts. Rewrite each as a multiplication equation. Show Part (a) on a number line.

a. $\frac{6}{11}$

b. $\frac{9}{4}$

c. $\frac{12}{8}$

d. $\frac{27}{10}$

3. Express each of the following as the sum of a whole number and a fraction. Show Parts (c) and (d) on number lines.

   a. $\frac{9}{5}$

   b. $\frac{7}{2}$

   c. $\frac{25}{7}$

   d. $\frac{21}{9}$

4. Natalie sawed five boards of equal length to make a stool. Each was 9 tenths of a meter long. What is the total length of the boards she sawed? Express your answer as the sum of a whole number and the remaining fractional units. Draw a number line to represent the problem.

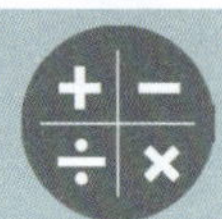

Topic B

Making Like Units Pictorially

5.NF.1, 5.NF.2

Focus Standards:	5.NF.1	Add and subtract fractions with unlike denominators (including mixed numbers) by replacing given fractions with equivalent fractions in such a way as to produce an equivalent sum or difference of fractions with like denominators. *For example, 2/3 + 5/4 = 8/12 + 15/12 = 23/12. (In general, a/b + c/d = (ad + bc)/bd.).*
	5.NF.2	Solve word problems involving addition and subtraction of fractions referring to the same whole, including cases of unlike denominators, e.g., by using visual fraction models or equations to represent the problem. Use benchmark fractions and number sense of fractions to estimate mentally and assess the reasonableness of answers. *For example, recognize an incorrect result 2/5 + 1/2 = 3/7, by observing that 3/7 < 1/2.*
Instructional Days:	5	
Coherence -Links from:	G4–M5	Fraction Equivalence, Ordering, and Operations
-Links to:	G5–M4	Multiplication and Division of Fractions and Decimal Fractions
	G6–M3	Rational Numbers

In Topic B, students use the familiar rectangular fraction model to add and subtract fractions with unlike denominators.

Students make like units for all addends or both minuend and subtrahend. First, they draw a wide rectangle and partition it with vertical lines as they would a tape diagram, representing the first fraction with a bracket and shading. They then partition a second congruent rectangle with horizontal lines to show the second fraction. Next, they partition both rectangles with matching lines to create like units.

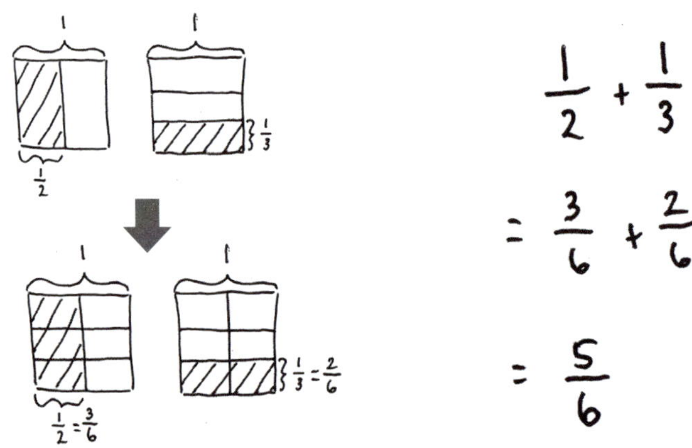

This strategy pictorially proves 3 sixths are equal to 1 half and 2 sixths are equal to 1 third. Students practice making these models extensively until they internalize the process of making like units. Students use the same systematic drawing for addition as they do for subtraction. In this manner, students are prepared to generalize with understanding to multiply the numerator and denominator by the same number. The topic closes with a lesson devoted to solving two-step word problems involving addition and subtraction of fractions.

A Teaching Sequence Toward Mastery of Making Like Units Pictorially	
Objective 1:	**Add fractions with unlike units using the strategy of creating equivalent fractions. (Lesson 3)**
Objective 2:	**Add fractions with sums between 1 and 2. (Lesson 4)**
Objective 3:	**Subtract fractions with unlike units using the strategy of creating equivalent fractions. (Lesson 5)**
Objective 4:	**Subtract fractions from numbers between 1 and 2. (Lesson 6)**
Objective 5:	**Solve two-step word problems. (Lesson 7)**

Lesson 3

Objective: Add fractions with unlike units using the strategy of creating equivalent fractions.

Suggested Lesson Structure

- Fluency Practice (12 minutes)
- Application Problem (5 minutes)
- Concept Development (33 minutes)
- Student Debrief (10 minutes)

Total Time (60 minutes)

Fluency Practice (12 minutes)

- Sprint: Find the Missing Numerator or Denominator **4.NF.1** (8 minutes)
- Adding Like Fractions **5.NF.1** (2 minutes)
- Rename the Fractions **5.NF.3** (2 minutes)

Sprint: Find the Missing Numerator or Denominator (8 minutes)

Materials: (S) Find the Missing Numerator or Denominator Sprint

Note: Students generate common equivalent fractions mentally and with automaticity (i.e., without performing the indicated multiplication).

Adding Like Fractions (2 minutes)

Note: This fluency activity reviews adding like units and lays the foundation for today's task of adding unlike units.

T: Let's add fractions mentally. Say answers as whole numbers when possible.

T: $\frac{1}{3} + \frac{1}{3} =$ ____?

S: $\frac{2}{3}$.

T: $\frac{1}{4} + \frac{1}{4} =$ ____?

S: $\frac{2}{4}$.

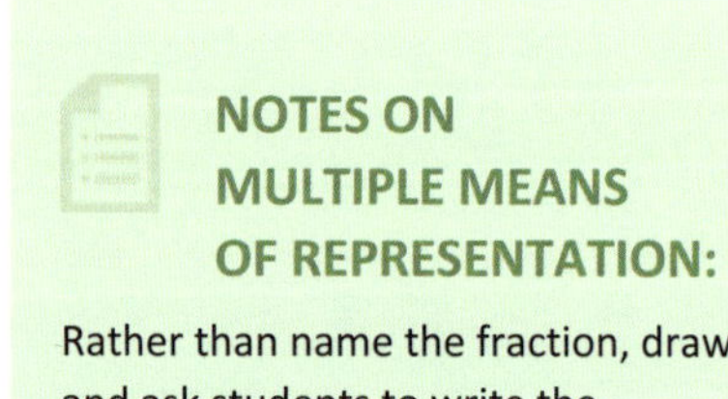

NOTES ON MULTIPLE MEANS OF REPRESENTATION:

Rather than name the fraction, draw it, and ask students to write the corresponding equation on their personal white boards. Use brackets to indicate the addends.

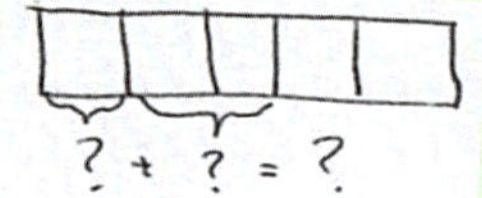

T: $\frac{1}{5} + \frac{2}{5} =$ ____?

S: $\frac{3}{5}$.

T: $\frac{3}{7} + \frac{4}{7} =$ ____?

S: 1.

T: $\frac{1}{4} + \frac{1}{3} + \frac{3}{4} + \frac{2}{3} =$ ____?

S: 2.

Continue and adjust to meet student needs. Use a variety of fraction combinations.

Rename the Fractions (2 minutes)

Materials: (S) Personal white board

Note: This fluency activity is a quick review of generating equivalent fractions, which students use as a strategy to add unlike units during today's Concept Development.

T: (Write $\frac{2}{4}$.) Rename the fraction by writing the largest units possible.

S: (Write $\frac{1}{2}$.)

T: (Write $\frac{3}{6}$.) Try this problem.

S: (Write $\frac{1}{2}$.)

Continue with the following possible sequence: $\frac{6}{12}$, $\frac{3}{9}$, $\frac{2}{6}$, $\frac{4}{6}$, $\frac{6}{9}$, $\frac{2}{8}$, $\frac{3}{12}$, $\frac{4}{16}$, $\frac{12}{16}$, $\frac{9}{12}$, and $\frac{6}{8}$.

Application Problem (5 minutes)

One ninth of the students in Mr. Beck's class list red as their favorite color. Twice as many students call blue their favorite, and three times as many students prefer pink. The rest name green as their favorite color. What fraction of the students say green or pink is their favorite color?

Extension: If 6 students call blue their favorite color, how many students are in Mr. Beck's class?

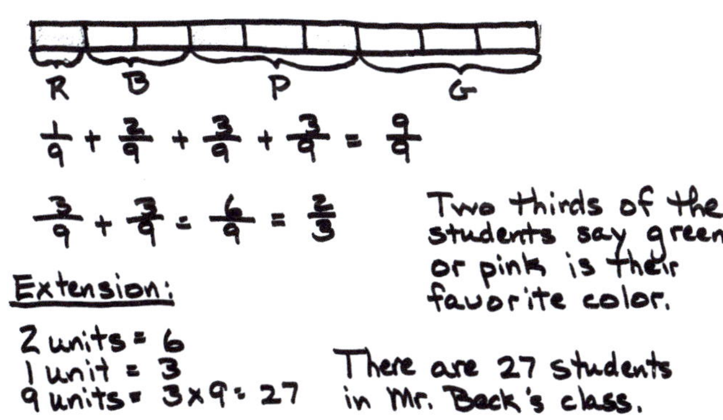

NOTES ON MULTIPLE MEANS OF ACTION AND EXPRESSION:

Students working above grade level may enjoy the challenge of an extension problem. If time permits, have one of the students model the extension problem on the board and share the solution with the class.

Concept Development (33 minutes)

Materials: (S) Personal white board, 2 pieces of $4\frac{1}{2}$" × $4\frac{1}{2}$" paper per student (depending on how the folding is completed before drawing the rectangular array model)

T: (Write 1 adult + 3 adults.) What is 1 adult plus 3 adults?
S: 4 adults.
T: 1 fifth plus 3 fifths?
S: 4 fifths.
T: We can add 1 fifth plus 3 fifths because the units are the same.

1 fifth + 3 fifths = 4 fifths.

$$\frac{1}{5}+\frac{3}{5}=\frac{4}{5}.$$

T: (Write 1 child + 3 adults.) What is 1 child plus 3 adults?
S: We can't add children and adults.
T: Why is that? Talk to your partner about that.
S: (Share.)
T: I heard Michael tell his partner that children and adults are not the same unit. We must replace unlike units with equivalent like units to add. What do children and adults have in common?
S: They are people.
T: Let's add people, not children and adults. Say the addition sentence with people.
S: 1 person + 3 people = 4 people.
T: Yes. What about 1 one plus 4 ones?
S: 5 ones.

NOTES ON MULTIPLE MEANS OF ENGAGEMENT:

Folding paper is a concrete strategy that helps build conceptual understanding. This helps ease the hardest part of using a rectangular fraction model—recognizing the original fractions once the horizontal lines are drawn. Help students see $\frac{1}{4}=\frac{2}{8}$ and $\frac{1}{2}=\frac{4}{8}$ by pointing and showing the following:

- $\frac{1}{4}=\frac{1}{8}+\frac{1}{8}$.
- $\frac{1}{2}=\frac{1}{8}+\frac{1}{8}+\frac{1}{8}+\frac{1}{8}$.

Problem 1: $\frac{1}{2}+\frac{1}{4}$

T: (Write $\frac{1}{2}+\frac{1}{4}$.) Can I add 1 half plus 1 fourth? Discuss with your partner. (Circulate and listen.) Pedro, could you share your thoughts?
S: I cannot add 1 half plus 1 fourth until the units are the same. We need to find like units.
T: Let's first make like units by folding paper. (Lead students through the process of folding as shown.)

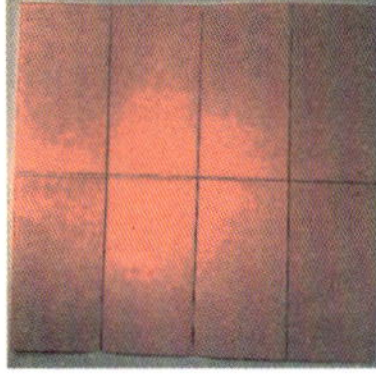
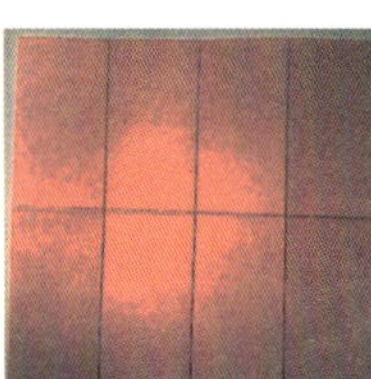

$$\frac{1}{2}+\frac{1}{4} \quad = \quad \frac{4}{8}+\frac{2}{8}$$

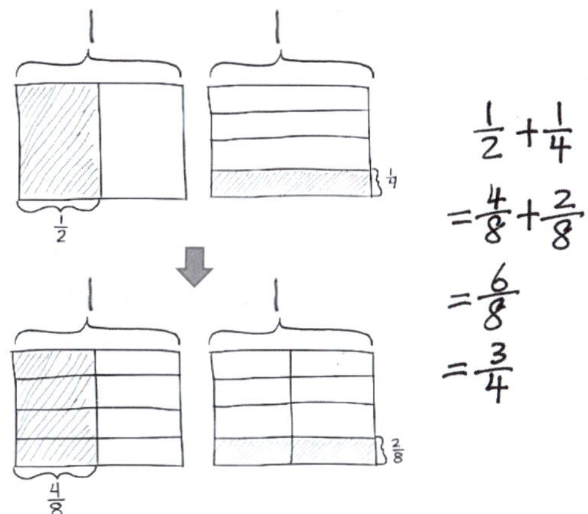

T: Now, let's make like units by drawing. (Draw a rectangular fraction model.) How many units will I have if I partition 1 whole into smaller units of one half each?

S: 2 units.

T: (Partition the rectangle vertically into 2 equal units.) One half tells me to select how many of the 2 units?

S: One.

T: Let's label our unit with $\frac{1}{2}$ and shade in one part. Now, let's draw another whole rectangle. How many equal parts do I divide this whole into to make fourths?

S: Four.

T: (Partition the rectangle horizontally into 4 equal units.) One fourth tells me to shade how many units?

S: One.

T: Let's label our unit with $\frac{1}{4}$ and shade in one part. Now, let's partition our 2 wholes into the same size units. (Draw horizontal lines on the $\frac{1}{2}$ model and 1 vertical line on the $\frac{1}{4}$ model.) What fractional unit have we made for each whole?

S: Eighths.

T: How many shaded units are in $\frac{1}{2}$?

S: Four.

T: That's right; we have 4 shaded units out of 8 total units. (Change the label from $\frac{1}{2}$ to $\frac{4}{8}$.) How many units are shaded on the $\frac{1}{4}$ model?

S: Two.

T: Yes, 2 shaded parts out of 8 total parts. (Change the label from $\frac{1}{4}$ to $\frac{2}{8}$.) Do our models show like units now?

S: Yes!

T: Say the addition sentence now using eighths as our common denominator, or common unit.

S: 4 eighths + 2 eighths = 6 eighths.

T: We can make larger units within $\frac{6}{8}$. Tell your partner how you might do that.

S: 6 and 8 can both be divided by 2. $6 \div 2 = 3$ and $8 \div 2 = 4$. The fraction is $\frac{3}{4}$. → We can make larger units of 2 each. 3 twos out of 4 twos. That's 3 out of 4 units or 3 fourths. → $\frac{6}{8}$ is partitioned into 6 out of 8 smaller units. It can be made into 3 out of 4 larger, equal pieces by grouping in 2s. 1 half + 1 fourth = 4 eighths + 2 eighths = 6 eighths = 3 fourths.

$$\frac{1}{2}+\frac{1}{4}=\frac{4}{8}+\frac{2}{8}=\frac{6}{8}=\frac{3}{4}.$$

Problem 2: $\frac{1}{3}+\frac{1}{2}$

In this problem, students can fold a paper again to transition into drawing or start directly with drawing. This is a simple problem involving two unit fractions, such as Problem 1. The primary purpose is to reinforce understanding of what is occurring to the units within a very simple context. Problem 3 moves forward to address a unit fraction plus a non-unit fraction.

T: Do our units get larger or smaller when we create like units? Talk to your partner.

S: The units get smaller. There are more units, and they are definitely getting smaller. → The units get smaller. It is the same amount of space but more parts. → We have to cut them up to make them the same size. → We can also think how 1 unit will become 6 units. That's what is happening to the half.

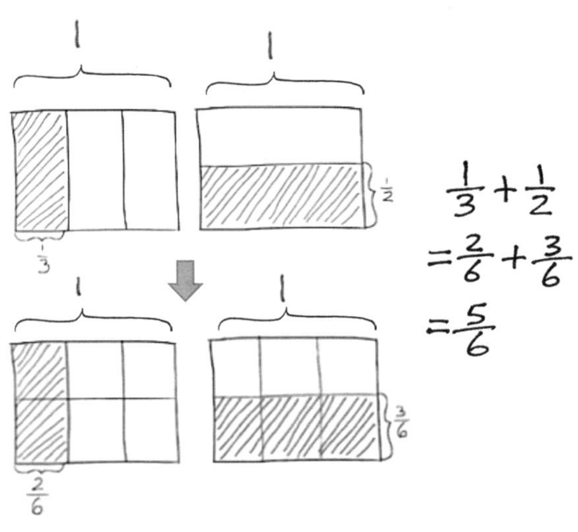

T: Let's draw a diagram to solve the problem and verify your thinking.

S: (Draw.)

T: Did the half become 3 smaller units and each third become 2 smaller units?

S: Yes!

T: Tell me the addition sentence.

S: 2 sixths + 3 sixths = 5 sixths.

$$\frac{1}{3}+\frac{1}{2}=\frac{2}{6}+\frac{3}{6}=\frac{5}{6}.$$

Problem 3: $\frac{2}{3}+\frac{1}{4}$

T: When we partition a rectangle into thirds, how many units do we have in all?

S: 3.

T: (Partition thirds vertically.) How many of those units are we shading?

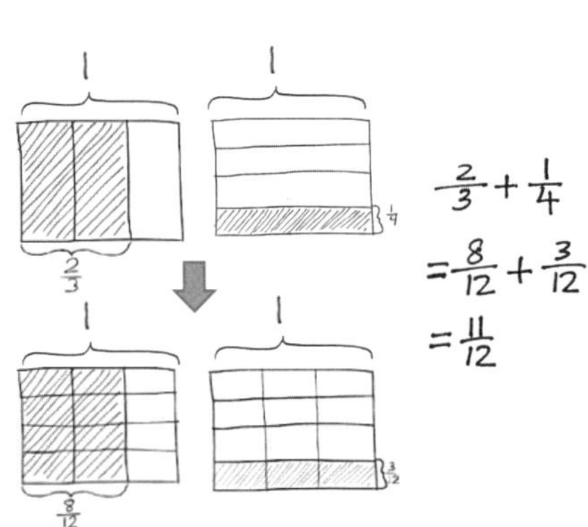

S: 2.

T: (Shade and label 2 thirds.) To show 1 fourth, how many units do we draw?

S: 4.

T: (Make a new rectangle of the same size and partition fourths horizontally.)

T: How many total units does this new rectangle have?

S: 4.

T: (Shade and label the new rectangle.)

T: Let's make these units the same size. (Partition the rectangles so the units are equal.)

T: What is the fractional value of 1 unit?

S: 1 twelfth.

T: How many twelfths are equal to 2 thirds?

S: 8 twelfths.

T: (Mark $\frac{8}{12}$ on the $\frac{2}{3}$ rectangle.) How many twelfths are equal to $\frac{1}{4}$?

S: 3 twelfths.

T: (Mark $\frac{3}{12}$ on the $\frac{1}{4}$ rectangle.) Say the addition sentence now using twelfths as our like unit or denominator.

S: 8 twelfths plus 3 twelfths equals 11 twelfths. $\frac{2}{3}+\frac{1}{4}=\frac{8}{12}+\frac{3}{12}=\frac{11}{12}$.

T: Read with me. 2 thirds + 1 fourth = 8 twelfths + 3 twelfths = 11 twelfths.

T: With your partner, review the process we used to solve $\frac{2}{3}+\frac{1}{4}$ step by step. Partner A goes first, and then Partner B. Draw an area model to show how you make equivalent fractions to add unlike units.

NOTES ON MULTIPLE MEANS OF REPRESENTATION:

For students who are confused about adding the parts together, have them cut out the parts of the second model and place them inside the first. For example, as shown in the drawings below, have them cut out the three one-twelfths and add them to the model with $\frac{8}{12}$, as if a puzzle. Have them speak the sentence, "8 twelfths plus 3 twelfths equals 11 twelfths." Repeat until students can visualize this process without the extra step.

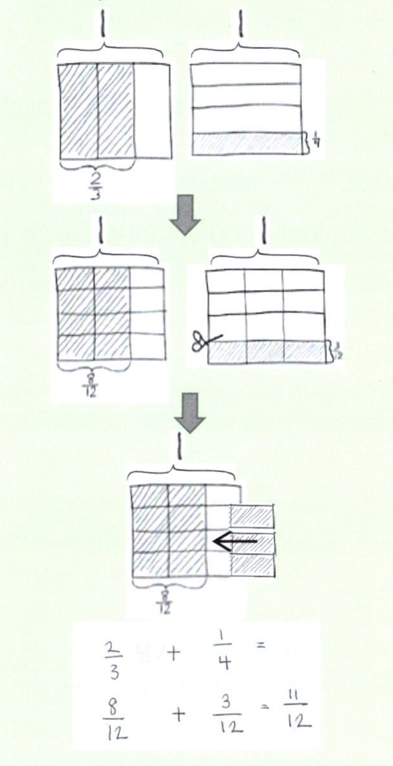

Problem 4: $\frac{2}{5}+\frac{2}{3}$

Note: This problem adds the complexity of finding the sum of two non-unit fractions, both with the numerator of 2. Working with fractions with common numerators invites healthy reflection on the size of fifths as compared to thirds. Students can reason that, while there are the same number of units (2), thirds are larger than fifths because the whole is broken into 3 parts instead of 5 parts. Therefore, there are more in each part. Additionally, it can be reasoned that 2 thirds is larger than 2 fifths because, when fifteenths are used for both, the number of units in 2 thirds (10) is more than the number used in 2 fifths (6).

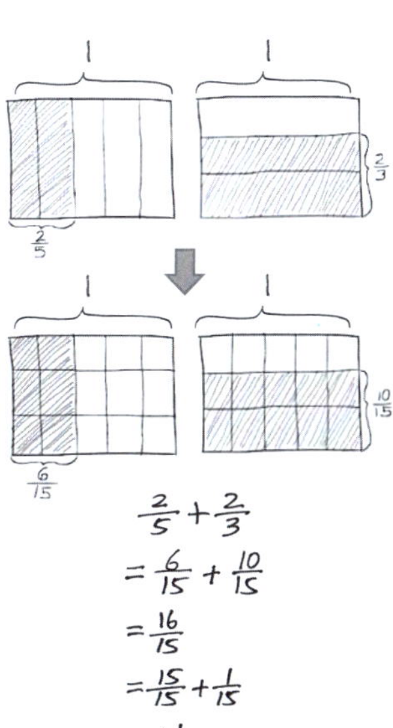

This problem also presents an opportunity to remind students about the importance of attending to precision (MP.6). When comparing fractions, care is taken to talk about the same whole amount as demonstrated by the rectangle. Such attention to precision also leads students to understand that 2 thirds of a cup is not larger than 2 fifths of a gallon.

Problem 5: $\frac{2}{7} + \frac{2}{3}$

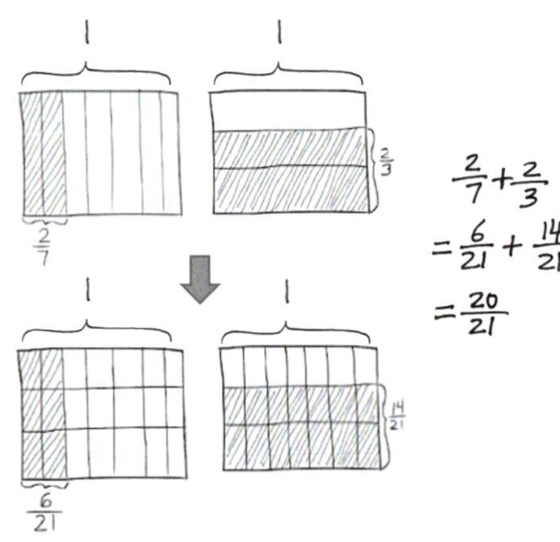

T: (Write $\frac{2}{7} + \frac{2}{3}$.) Work with your partner to solve this problem.

S: (Work.) 2 sevenths + 2 thirds = 6 twenty-firsts + 14 twenty-firsts = 20 twenty-firsts.

$$\frac{2}{7} + \frac{2}{3} = \frac{6}{21} + \frac{14}{21} = \frac{20}{21}.$$

Problem Set (10 minutes)

Students should do their personal best to complete the Problem Set within the allotted 10 minutes. For some classes, it may be appropriate to modify the assignment by specifying which problems they work on first. Some problems do not specify a method for solving. Students should solve these problems using the RDW approach used for Application Problems.

Student Debrief (10 minutes)

Lesson Objective: Add fractions with unlike units using the strategy of creating equivalent fractions.

The Student Debrief is intended to invite reflection and active processing of the total lesson experience.

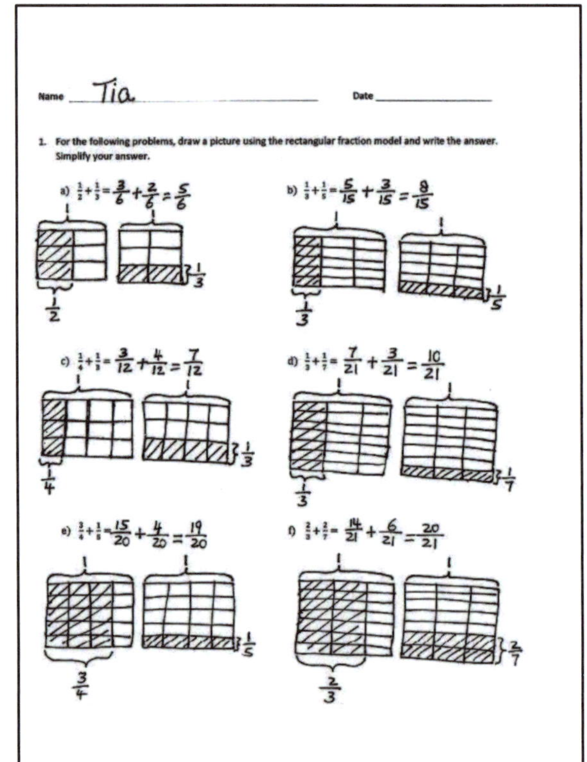

Invite students to review their solutions for the Problem Set. They should check work by comparing answers with a partner before going over answers as a class. Look for misconceptions or misunderstandings that can be addressed in the Debrief. Guide students in a conversation to debrief the Problem Set and process the lesson.

T: For one minute, go over the answers to Problem 1 with your partner. Don't change your work.

S: (Work together.)

T: Now, let's correct errors together. I will say the addition problem; you will say the answer. Problem 1(a). 1 half plus 1 third is...?

S: 5 sixths.

Continue with Problem 1(b–f). Then, give students about 2 minutes to correct their errors.

T: Analyze the following problems. How are they related?

- Problem 1 (a) and (b)
- Problem 1 (a) and (c)

- Problem 1 (b) and (d)
- Problem 1 (d) and (f)

S: (Discuss.)

T: Steven noticed something about Problem 1 (a) and (b). Please share.

S: The answer to Problem 1(b) is smaller than Problem 1(a) since you are adding only $\frac{1}{5}$ to $\frac{1}{2}$. Both answers are less than 1, but Problem 1(a) is much closer to 1. Problem 1(b) is really close to $\frac{1}{2}$ because $\frac{8}{16}$ would be $\frac{1}{2}$.

MP.7

T: Kara, can you share what you noticed about Problem 1 (d) and (f)?

S: I noticed that both problems used thirds and sevenths. But the numerators in Problem 1(d) were 1, and the numerators in Problem 1(f) were 2. Since the numerators doubled, the answer doubled from 10 twenty-firsts to 20 twenty-firsts.

T: I am glad to hear you are able to point out relationships between different problems.

T: Share with your partner what you learned how to do today.

S: (Share.)

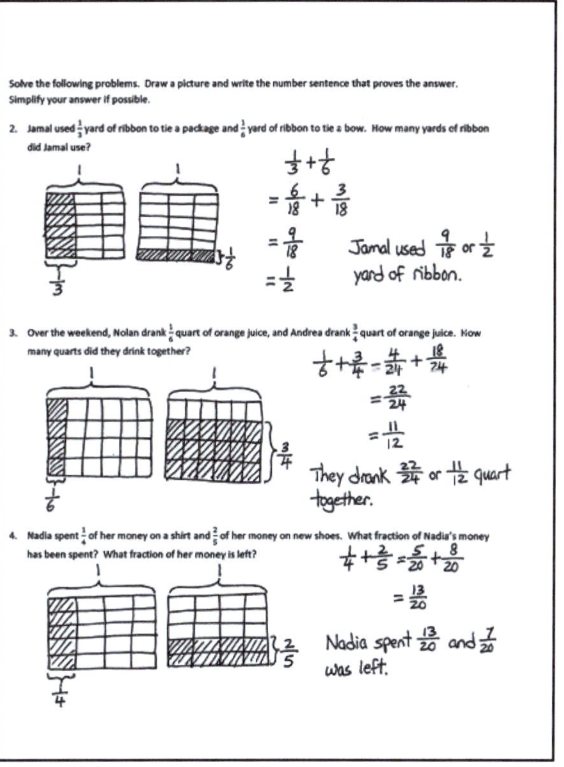

T: (Help students name the objective: We learned how to add fractions that have unlike units using a rectangular fraction model to create like units.)

Exit Ticket (3 minutes)

After the Student Debrief, instruct students to complete the Exit Ticket. A review of their work will help with assessing students' understanding of the concepts that were presented in today's lesson and planning more effectively for future lessons. The questions may be read aloud to the students.

A

Number Correct: _______

Find the Missing Numerator or Denominator

1.	$\frac{1}{2} = \frac{}{4}$	
2.	$\frac{1}{5} = \frac{2}{}$	
3.	$\frac{2}{5} = \frac{}{10}$	
4.	$\frac{3}{5} = \frac{}{10}$	
5.	$\frac{4}{5} = \frac{}{10}$	
6.	$\frac{1}{3} = \frac{2}{}$	
7.	$\frac{2}{3} = \frac{}{6}$	
8.	$\frac{1}{3} = \frac{3}{}$	
9.	$\frac{2}{3} = \frac{}{9}$	
10.	$\frac{1}{4} = \frac{}{8}$	
11.	$\frac{3}{4} = \frac{}{8}$	
12.	$\frac{1}{4} = \frac{3}{}$	
13.	$\frac{3}{4} = \frac{9}{}$	
14.	$\frac{2}{4} = \frac{}{2}$	
15.	$\frac{2}{6} = \frac{1}{}$	
16.	$\frac{2}{10} = \frac{1}{}$	
17.	$\frac{4}{10} = \frac{}{5}$	
18.	$\frac{8}{10} = \frac{}{5}$	
19.	$\frac{3}{9} = \frac{}{3}$	
20.	$\frac{6}{9} = \frac{}{3}$	
21.	$\frac{3}{12} = \frac{1}{}$	
22.	$\frac{9}{12} = \frac{}{4}$	
23.	$\frac{1}{3} = \frac{}{12}$	
24.	$\frac{2}{3} = \frac{}{12}$	
25.	$\frac{8}{12} = \frac{}{3}$	
26.	$\frac{12}{16} = \frac{3}{}$	
27.	$\frac{3}{5} = \frac{}{25}$	
28.	$\frac{4}{5} = \frac{28}{}$	
29.	$\frac{18}{24} = \frac{3}{}$	
30.	$\frac{24}{30} = \frac{}{5}$	
31.	$\frac{5}{6} = \frac{35}{}$	
32.	$\frac{56}{63} = \frac{}{9}$	
33.	$\frac{64}{72} = \frac{8}{}$	
34.	$\frac{5}{8} = \frac{}{64}$	
35.	$\frac{5}{6} = \frac{45}{}$	
36.	$\frac{45}{81} = \frac{}{9}$	
37.	$\frac{6}{7} = \frac{48}{}$	
38.	$\frac{36}{81} = \frac{}{9}$	
39.	$\frac{8}{56} = \frac{1}{}$	
40.	$\frac{35}{63} = \frac{5}{}$	
41.	$\frac{1}{6} = \frac{12}{}$	
42.	$\frac{3}{7} = \frac{36}{}$	
43.	$\frac{48}{60} = \frac{4}{}$	
44.	$\frac{72}{84} = \frac{}{7}$	

B

Number Correct: _______

Improvement: _______

Find the Missing Numerator or Denominator

1.	$\frac{1}{5} = \frac{2}{\ }$	
2.	$\frac{2}{5} = \frac{\ }{10}$	
3.	$\frac{3}{5} = \frac{\ }{10}$	
4.	$\frac{4}{5} = \frac{\ }{10}$	
5.	$\frac{1}{3} = \frac{2}{\ }$	
6.	$\frac{1}{3} = \frac{\ }{6}$	
7.	$\frac{2}{3} = \frac{4}{\ }$	
8.	$\frac{1}{3} = \frac{\ }{9}$	
9.	$\frac{2}{3} = \frac{6}{\ }$	
10.	$\frac{1}{4} = \frac{2}{\ }$	
11.	$\frac{3}{4} = \frac{6}{\ }$	
12.	$\frac{1}{4} = \frac{\ }{12}$	
13.	$\frac{3}{4} = \frac{\ }{12}$	
14.	$\frac{2}{4} = \frac{1}{\ }$	
15.	$\frac{2}{6} = \frac{\ }{3}$	
16.	$\frac{2}{10} = \frac{\ }{5}$	
17.	$\frac{4}{10} = \frac{2}{\ }$	
18.	$\frac{8}{10} = \frac{4}{\ }$	
19.	$\frac{3}{9} = \frac{1}{\ }$	
20.	$\frac{6}{9} = \frac{2}{\ }$	
21.	$\frac{1}{4} = \frac{\ }{12}$	
22.	$\frac{9}{12} = \frac{3}{\ }$	
23.	$\frac{1}{3} = \frac{4}{\ }$	
24.	$\frac{2}{3} = \frac{8}{\ }$	
25.	$\frac{8}{12} = \frac{2}{\ }$	
26.	$\frac{12}{16} = \frac{\ }{4}$	
27.	$\frac{3}{5} = \frac{15}{\ }$	
28.	$\frac{4}{5} = \frac{\ }{35}$	
29.	$\frac{18}{24} = \frac{\ }{4}$	
30.	$\frac{24}{30} = \frac{4}{\ }$	
31.	$\frac{5}{6} = \frac{\ }{42}$	
32.	$\frac{56}{63} = \frac{8}{\ }$	
33.	$\frac{64}{72} = \frac{\ }{9}$	
34.	$\frac{5}{8} = \frac{40}{\ }$	
35.	$\frac{5}{6} = \frac{\ }{54}$	
36.	$\frac{45}{81} = \frac{5}{\ }$	
37.	$\frac{6}{7} = \frac{\ }{56}$	
38.	$\frac{36}{81} = \frac{4}{\ }$	
39.	$\frac{8}{56} = \frac{\ }{7}$	
40.	$\frac{35}{63} = \frac{\ }{9}$	
41.	$\frac{1}{6} = \frac{\ }{72}$	
42.	$\frac{3}{7} = \frac{\ }{84}$	
43.	$\frac{48}{60} = \frac{\ }{5}$	
44.	$\frac{72}{84} = \frac{6}{\ }$	

Name ______________________________ Date ______________

1. Draw a rectangular fraction model to find the sum. Simplify your answer, if possible.

a. $\frac{1}{2} + \frac{1}{3} =$

b. $\frac{1}{3} + \frac{1}{5} =$

c. $\frac{1}{4} + \frac{1}{3} =$

d. $\frac{1}{3} + \frac{1}{7} =$

e. $\frac{3}{4}+\frac{1}{5}=$

f. $\frac{2}{3}+\frac{2}{7}=$

Solve the following problems. Draw a picture and write the number sentence that proves the answer. Simplify your answer, if possible.

2. Jamal used $\frac{1}{3}$ yard of ribbon to tie a package and $\frac{1}{6}$ yard of ribbon to tie a bow. How many yards of ribbon did Jamal use?

3. Over the weekend, Nolan drank $\frac{1}{6}$ quart of orange juice, and Andrea drank $\frac{3}{4}$ quart of orange juice. How many quarts did they drink together?

4. Nadia spent $\frac{1}{4}$ of her money on a shirt and $\frac{2}{5}$ of her money on new shoes. What fraction of Nadia's money has been spent? What fraction of her money is left?

Name ______________________________ Date ________________

Solve by drawing the rectangular fraction model.

1. $\frac{1}{2} + \frac{1}{5} =$

2. In one hour, Ed used $\frac{2}{5}$ of the time to complete his homework and $\frac{1}{4}$ of the time to check his email. How much time did he spend completing homework and checking email? Write your answer as a fraction. (Extension: Write the answer in minutes.)

Name ____________________________________ Date ____________________

1. Draw a rectangular fraction model to find the sum. Simplify your answer, if possible.

a. $\frac{1}{4}+\frac{1}{3}=$

b. $\frac{1}{4}+\frac{1}{5}=$

c. $\frac{1}{4}+\frac{1}{6}=$

d. $\frac{1}{5}+\frac{1}{9}=$

e. $\frac{1}{4} + \frac{2}{5} =$

f. $\frac{3}{5} + \frac{3}{7} =$

Solve the following problems. Draw a picture, and write the number sentence that proves the answer. Simplify your answer, if possible.

2. Rajesh jogged $\frac{3}{4}$ mile and then walked $\frac{1}{6}$ mile to cool down. How far did he travel?

3. Cynthia completed $\frac{2}{3}$ of the items on her to-do list in the morning and finished $\frac{1}{8}$ of the items during her lunch break. What fraction of her to-do list is finished by the end of her lunch break? (Extension: What fraction of her to-do list does she still have to do after lunch?)

4. Sam read $\frac{2}{5}$ of her book over the weekend and $\frac{1}{6}$ of it on Monday. What fraction of the book has she read? What fraction of the book is left?

Lesson 4

Objective: Add fractions with sums between 1 and 2.

Suggested Lesson Structure

Fluency Practice	(8 minutes)
Application Problems	(7 minutes)
Concept Development	(35 minutes)
Student Debrief	(10 minutes)
Total Time	**(60 minutes)**

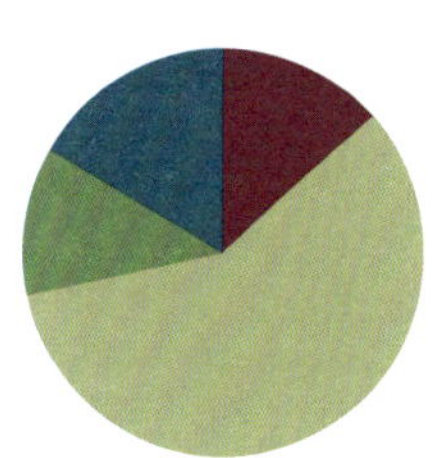

Fluency Practice (8 minutes)

- Adding Fractions to Make One Whole **4.NF.3a** (4 minutes)
- Skip-Counting by $\frac{1}{3}$ Yard **5.MD.1** (4 minutes)

Adding Fractions to Make One Whole (4 minutes)

Note: This fluency activity is a quick mental exercise of part–part–whole understanding as it relates to fractions.

T: I will name a fraction. You say a fraction with the same denominator so that together our fractions add up to 1 whole. For example, if I say 1 third, you say 2 thirds. $\frac{1}{3} + \frac{2}{3} = \frac{3}{3}$ or 1 whole. Say your answer at the signal.

T: 1 fourth? (Signal.)

S: 3 fourths.

T: 1 fifth? (Signal.)

S: 4 fifths.

T: 2 tenths? (Signal.)

S: 8 tenths.

Continue with the following possible sequence: $\frac{1}{3}$, $\frac{3}{5}$, $\frac{1}{2}$, $\frac{5}{10}$, $\frac{6}{7}$, and $\frac{3}{8}$.

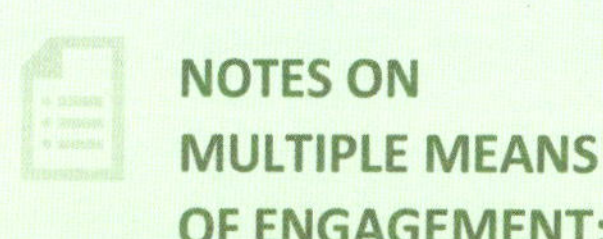

NOTES ON MULTIPLE MEANS OF ENGAGEMENT:

Depending on the group of students, consider supporting them visually by making fraction cards that show circles divided into fourths, fifths, tenths, etc. Flash the corresponding card while naming the fraction. English language learners have a visual support to accompany language, and students working below grade level can see how many more to make one whole.

Skip-Counting by $\frac{1}{3}$ Yard (4 minutes)

Note: This skip-counting fluency activity prepares students for success with addition and subtraction of fractions between 1 and 2.

T: Let's count by $\frac{1}{3}$ yard. (Rhythmically point up until a change is desired. Show a closed hand, and then point down. Continue, mixing it up.)

S: $\frac{1}{3}$ yard, $\frac{2}{3}$ yard, 1 yard, (stop), $\frac{2}{3}$ yard, (stop), 1 yard, $1\frac{1}{3}$ yards, $1\frac{2}{3}$ yards, 2 yards, (stop), $1\frac{2}{3}$ yards, $1\frac{1}{3}$ yards, 1 yard, (stop).

Continue the sequence going up to, and beyond, 3 yards, paying careful attention when crossing over whole number units.

NOTES ON MULTIPLE MEANS OF ENGAGEMENT:

Periodically challenge students working above grade level to rename each fraction of a yard as a number of feet.

Application Problem (7 minutes)

Leslie has 1 liter of milk in her refrigerator to drink today. She drank $\frac{1}{2}$ liter of milk for breakfast and $\frac{2}{5}$ liter of milk for dinner. How much of a liter did Leslie drink during breakfast and dinner?

(Extension: How much of a liter of milk does Leslie have left to drink with her dessert? Give your answer as a fraction of liters and as a decimal.)

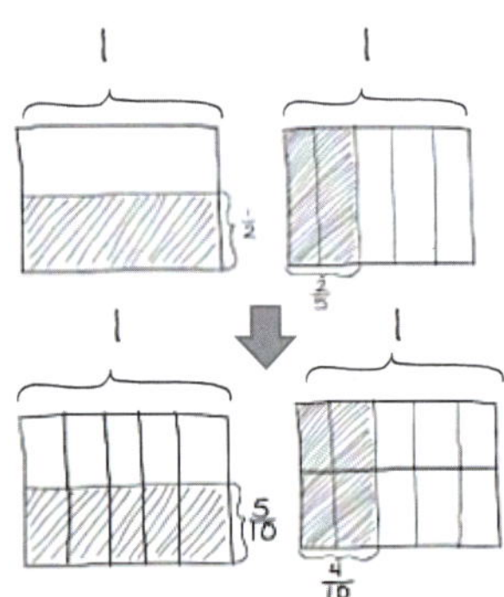

$\frac{1}{2} + \frac{2}{5} = \frac{5}{10} + \frac{4}{10} = \frac{9}{10}$

Leslie drank $\frac{9}{10}$ L, or 0.9 L.

$1 - \frac{9}{10} = \frac{1}{10}$

She had $\frac{1}{10}$ L left, or 0.1 L.

T: Let's read the problem together.

S: (Read chorally.)

T: What is our whole?

S: 1 liter.

T: Tell your partner how you might solve this problem.

S: (Discuss.)

T: (Select a student to draw a model for this problem.) I see that Joe has a great model to help us solve this problem. Joe, please draw your picture for us on the board.

S: (Draw.)

NOTES ON MULTIPLE MEANS OF ACTION AND EXPRESSION:

It can be helpful to English language learners to have others model speech to describe the models they draw. If appropriate, select an English language learner to help make the drawing for the class.

T: Thank you, Joe. Let's say an addition expression that represents this word problem.

S: 2 fifths plus 1 half.

T: Why can't we add these two fractions?

S: They are different. → They have different denominators. → The units are different. We must find a like unit between fifths and halves. → We can use equal fractions to add them. → The fractions will look different, but they will still be the same amount.

T: Joe found like units from his drawing. How many units are inside his rectangle?

S: 10.

T: That means we will use 10 as our denominator, or our named unit, to solve this problem. Say your addition sentence now using tenths.

S: 4 tenths plus 5 tenths equals 9 tenths.

T: Good. Please say a sentence to your partner about how much milk Leslie drank for breakfast and dinner.

S: Leslie drank $\frac{9}{10}$ liter of milk for breakfast and dinner.

T: With words, how would you write 9 tenths as a decimal?

S: Zero point nine.

T: Now, we need to solve the extension question. How much milk will Leslie have available for dessert? Tell your partner how you solved this.

S: I know Leslie drank $\frac{9}{10}$ liter of milk so far. I know she has 1 whole liter, which is 10 tenths. 9 tenths plus 1 tenth equals 10 tenths, so Leslie has 1 tenth liter of milk for her dessert.

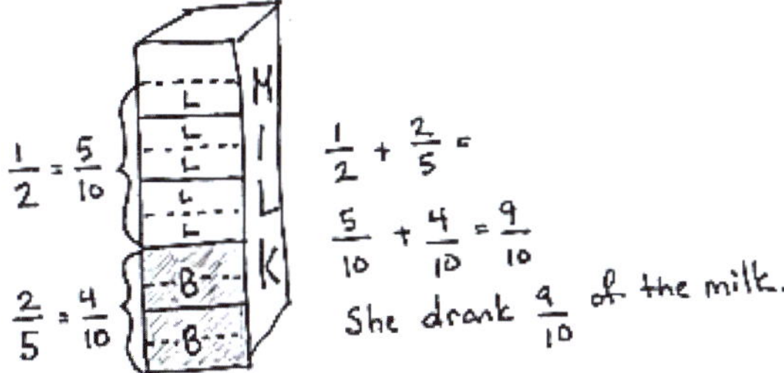

Note: Students solve this Application Problem involving addition of fractions with unlike denominators, using visual models as learned in Lesson 3.

Concept Development (35 minutes)

Materials: (S) Personal white board

Problem 1: a. $\frac{1}{3}+\frac{1}{4}$ **b.** $\frac{1}{2}+\frac{3}{4}$

T: (Write Problem 1(a) on the board.) When you see this problem, can you estimate the answer? Will it be more or less than 1? Talk with your partner.

S: The answer is less than 1 because $\frac{1}{3}$ and $\frac{1}{4}$ are both less than $\frac{1}{2}$. So, if two fractions that are each less than $\frac{1}{2}$ are added together, they will add up to a fraction less than 1 whole.

T: (Write Problem 1(b) on the board.) Now, look at this problem. Estimate the answer.

S: (Discuss.)

T: I overheard Camden say the answer will be more than 1 whole. Can you explain why you think so?

S $\frac{3}{4}$ is more than 1 half and it's added to 1 half; we will have a sum greater than 1 whole.

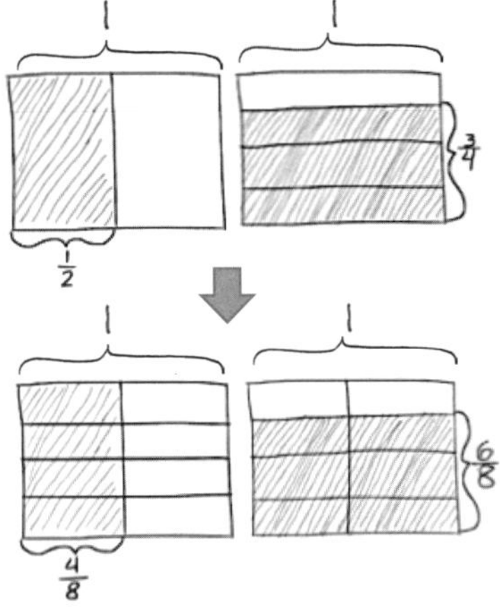

T: What stops us from simply adding?

S: The units are not the same.

T: (Draw two rectangular fraction models.) How many parts do I need to draw for 1 half?

S: 2.

T: (Partition one rectangle into 2 units.) How many parts should I shade and label to show 1 half?

S: 1.

T: Just like the previous lesson, we label our picture with $\frac{1}{2}$. Now, let's partition this other rectangle horizontally. How many rows to show fourths?

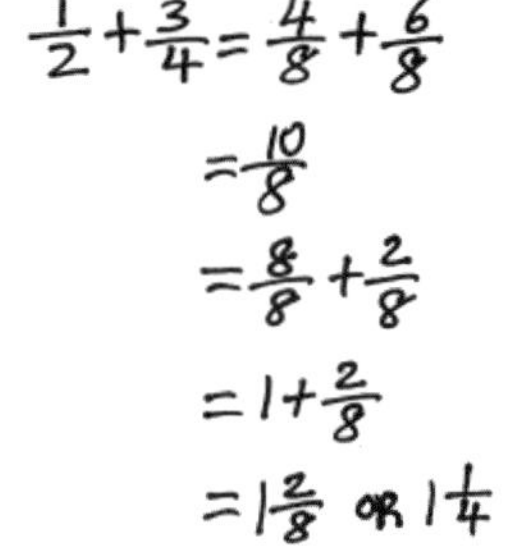

S: 4.

T: How many rows do we shade to represent 3 fourths?

S: 3.

T: We bracket 3 fourths of this rectangle. Now, let's partition both wholes into units of the same size. How many parts do we need in each rectangle to make the units the same size?

S: 8.

T: (Partition the models.) What is the fractional value of one unit now?

S: 1 eighth.

T: Eighths is the like unit or common denominator. We can decompose $\frac{1}{2}$ into eighths. How many eighths are equal to 1 half? (Point to the 4 boxes bracketed by $\frac{1}{2}$.)

S: 4 eighths.

T: How many eighths are the same as $\frac{3}{4}$? (Point out the 6 boxes bracketed by $\frac{3}{4}$.)

S: 6 eighths.

T: Say the addition sentence now using eighths as our common denominator.

S: 4 eighths plus 6 eighths equals 10 eighths. 1 half + 3 fourths = 4 eighths + 6 eighths = 10 eighths.

T: Good. What is unusual about our answer 10 eighths? Tell your partner.

S: The answer has a numerator larger than its denominator. We can write it as a mixed number instead. → Ten eighths is more than 1 whole.

T: How many eighths make 1 whole?

S: 8 eighths.

T: 8 eighths plus what equals 10 eighths?

S: 2 eighths.

T: Did anyone use another unit to express your answer?

S: I used fourths. I know that eighths are half as large as fourths. So, 2 eighths is the same amount as 1 fourth.

T: Can you share your answer, the sum, with us?

S: 1 and 1 fourth.

Problem 2: $\frac{4}{5}+\frac{1}{2}$

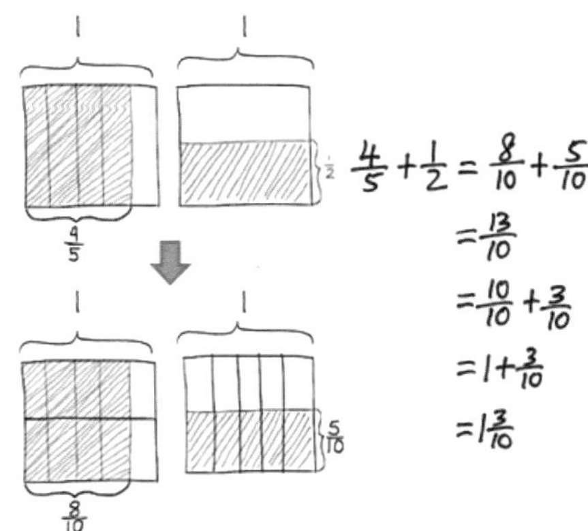

T: (Write $\frac{4}{5}+\frac{1}{2}$.) Solve this problem.

S: (Solve.)

T: Share with your partner how to express 13 tenths as a mixed number.

S: 10 tenths plus 3 tenths equals 13 tenths. 10 tenths makes a whole and 3 tenths is left over. The sum is 1 and $\frac{3}{10}$.

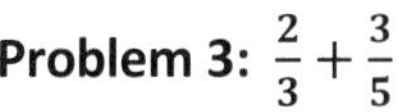

Problem 3: $\frac{2}{3}+\frac{3}{5}$

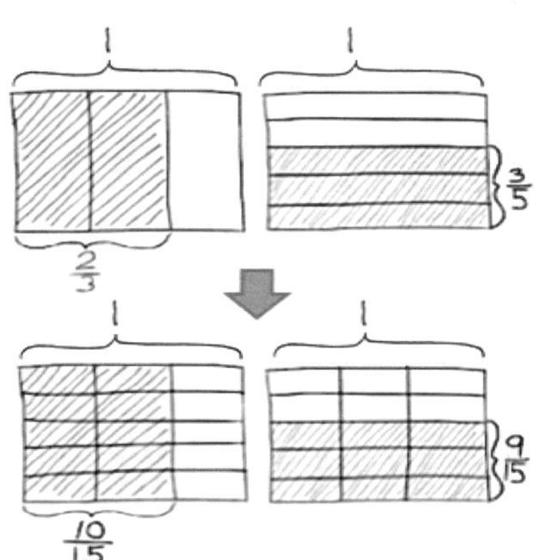

$$\frac{2}{3}+\frac{3}{5}=\frac{10}{15}+\frac{9}{15}$$
$$=\frac{19}{15}$$
$$=\frac{15}{15}+\frac{4}{15}$$
$$=1+\frac{4}{15}$$
$$=1\frac{4}{15}$$

T: (Write $\frac{2}{3}+\frac{3}{5}$.) Let's try another. Both addends have numerators greater than one, so make sure your brackets are clear. Draw the model you will use to solve.

S: (Draw.)

T: Discuss with your partner what you bracketed and why.
I'll walk around to see how it's going. (Allow one minute for students to discuss.)

T: What's another way to express $\frac{19}{15}$?

S: Write it as a mixed number.

T: Do that now individually. (Allow 1 minute to work.) Compare your work with your partner. What is the sum of 2 thirds plus 3 fifths?

S: 1 and 4 fifteenths.

Problem 4: $\frac{3}{8} + \frac{2}{3}$

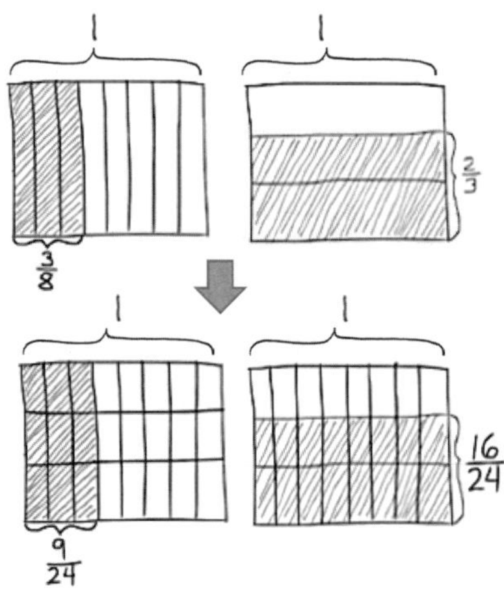

T: (Write $\frac{3}{8} + \frac{2}{3}$.) Try to solve this problem on your own. Draw a rectangular fraction model, and write a number sentence. Once everyone is finished, we will check your work.

S: (Work.)

T: What's the like unit or common denominator for eighths and thirds?

S: Twenty-fourths.

T: Say your addition sentence using twenty-fourths.

S: 9 twenty-fourths plus 16 twenty-fourths equals 25 twenty-fourths.

T: How can $\frac{25}{24}$ be changed to a mixed number?

S: 25 twenty-fourths = 24 twenty-fourths + 1 twenty-fourth.

T: What's another way to express $\frac{24}{24}$?

S: 1 whole.

T: Say the sum of 3 eighths plus 2 thirds.

S: 1 and 1 twenty-fourth.

Problem Set (10 minutes)

Students should do their personal best to complete the Problem Set within the allotted 10 minutes. For some classes, it may be appropriate to modify the assignment by specifying which problems they work on first. Some problems do not specify a method for solving. Students should solve these problems using the RDW approach used for Application Problems.

Student Debrief (10 minutes)

Lesson Objective: Add fractions with sums between 1 and 2.

The Student Debrief is intended to invite reflection and active processing of the total lesson experience.

Invite students to review their solutions for the Problem Set. They should check work by comparing answers with a partner before going over answers as a class. Look for misconceptions or misunderstandings that can be addressed in the Debrief. Guide students in a conversation to debrief the Problem Set and process the lesson.

T: Have your Problem Set ready to correct. I will say the addition expression. You say the sum as a mixed number. Problem 1(a), 2 thirds plus 1 half...?

S: 1 and 1 sixth.

Continue in this way for the entire Problem Set.

G5-M3-TE-B3-1.3.1-1.2016

T: I am going to give you 2 minutes to talk to your partner about any relationships you noticed on today's Problem Set. Be specific.

Allow for students to discuss. Then, proceed with a similar conversation to the one below.

T: Ryan, I heard you talking about Problem 1 (a) and (c). Can you share what you found with the class?

S: I saw that both problems used 1 half. So, I compared the second fraction and saw that they used $\frac{2}{3}$ in Problem 1(a) and $\frac{3}{5}$ in Problem 1(c). I remember from comparing fractions last year that $\frac{2}{3}$ is greater than $\frac{3}{5}$. It is really close. $\frac{2}{3}$ is $\frac{10}{15}$ and $\frac{3}{5}$ is $\frac{9}{15}$. So, the answers for Problem 1(a) and Problem 1(c) also show that Problem 1(a) is greater than Problem 1(c) because Problem 1(a) adds $\frac{2}{3}$.

T: Thank you, Ryan. Can someone else share, please?

S: I noticed that every single fraction on this Problem Set is greater than or equal to one half. That means when I add two fractions that are greater than one half together, my answer will be greater than 1. That also means that I will have to change my answer to a mixed number.

T: Thank you. Now, I will give you 1 minute to look at Jacqueline's work. What tool did she use to convert her fractions greater than 1 to mixed numbers?

S: Number bonds!

T: Turn and talk to your neighbor briefly about what you observe about her use of number bonds and how that compared with your conversion method.

T: What tool did you use to convert your fractions into like units?

S: The rectangle model.

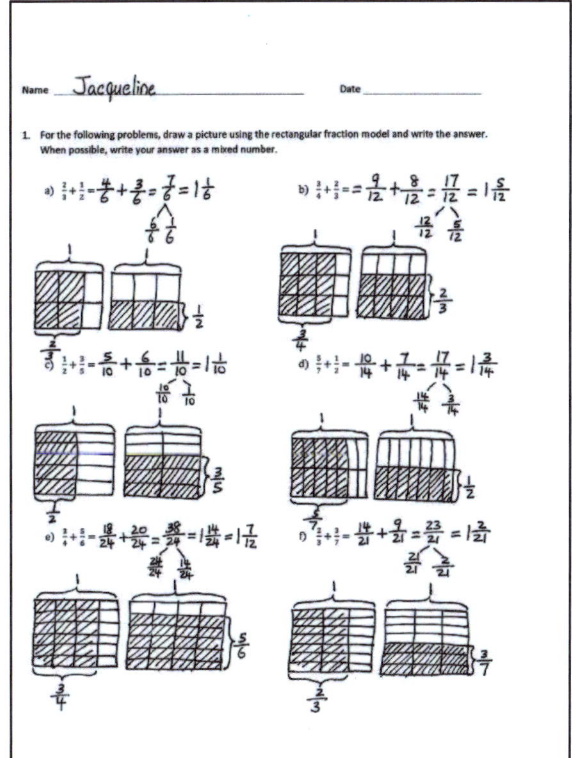
Name Jacqueline Date

1. For the following problems, draw a picture using the rectangular fraction model and write the answer. When possible, write your answer as a mixed number.

a) $\frac{2}{3}+\frac{1}{2}=\frac{4}{6}+\frac{3}{6}=\frac{7}{6}=1\frac{1}{6}$

b) $\frac{3}{4}+\frac{2}{3}==\frac{9}{12}+\frac{8}{12}=\frac{17}{12}=1\frac{5}{12}$

c) $\frac{1}{2}+\frac{3}{5}=\frac{5}{10}+\frac{6}{10}=\frac{11}{10}=1\frac{1}{10}$

d) $\frac{5}{7}+\frac{1}{2}=\frac{10}{14}+\frac{7}{14}=\frac{17}{14}=1\frac{3}{14}$

e) $\frac{3}{4}+\frac{5}{6}=\frac{18}{24}+\frac{20}{24}=\frac{38}{24}=1\frac{14}{24}=1\frac{7}{12}$

f) $\frac{2}{3}+\frac{3}{7}=\frac{14}{21}+\frac{9}{21}=\frac{23}{21}=1\frac{2}{21}$

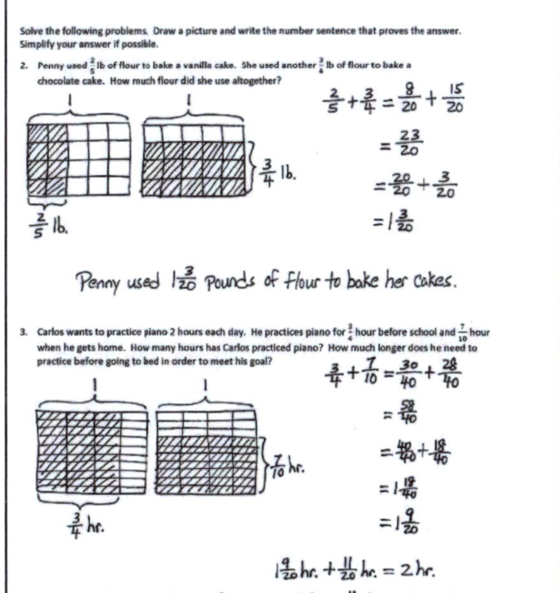
Solve the following problems. Draw a picture and write the number sentence that proves the answer. Simplify your answer if possible.

2. Penny used $\frac{2}{5}$ lb of flour to bake a vanilla cake. She used another $\frac{3}{4}$ lb of flour to bake a chocolate cake. How much flour did she use altogether?

$\frac{2}{5}+\frac{3}{4}=\frac{8}{20}+\frac{15}{20}=\frac{23}{20}=\frac{20}{20}+\frac{3}{20}=1\frac{3}{20}$

Penny used $1\frac{3}{20}$ pounds of flour to bake her cakes.

3. Carlos wants to practice piano 2 hours each day. He practices piano for $\frac{3}{4}$ hour before school and $\frac{7}{10}$ hour when he gets home. How many hours has Carlos practiced piano? How much longer does he need to practice before going to bed in order to meet his goal?

$\frac{3}{4}+\frac{7}{10}=\frac{30}{40}+\frac{28}{40}=\frac{58}{40}=\frac{40}{40}+\frac{18}{40}=1\frac{18}{40}=1\frac{9}{20}$

$1\frac{9}{20}$ hr. $+\frac{11}{20}$ hr. $=2$ hr.

Carlos has practiced $1\frac{9}{20}$ hours and has $\frac{11}{20}$ hour to go to make his goal.

T: (After students share.) How does this work today relate to our work yesterday?

S: Again, we took larger units and broke them into smaller equal units to find like denominators. → Yesterday, all of our answers were less than 1 whole. Today, we realized we could use the model when the sum is greater than 1. → Our model doesn't show the sum of the units. It just shows us the number of units that we must use to add. → Yeah, that meant we didn't have to draw a whole other rectangle. → I get it better today than yesterday. Now, I really see what is happening.

T: Show me your learning on the Exit Ticket!

Exit Ticket (3 minutes)

After the Student Debrief, instruct students to complete the Exit Ticket. A review of their work will help with assessing students' understanding of the concepts that were presented in today's lesson and planning more effectively for future lessons. The questions may be read aloud to the students.

Name ________________________________ Date ________________

1. For the following problems, draw a picture using the rectangular fraction model and write the answer. When possible, write your answer as a mixed number.

a. $\frac{2}{3}+\frac{1}{2}=$

b. $\frac{3}{4}+\frac{2}{3}=$

c. $\frac{1}{2}+\frac{3}{5}=$

d. $\frac{5}{7}+\frac{1}{2}=$

e. $\frac{3}{4}+\frac{5}{6}=$

f. $\frac{2}{3}+\frac{3}{7}=$

Solve the following problems. Draw a picture, and write the number sentence that proves the answer. Simplify your answer, if possible.

2. Penny used $\frac{2}{5}$ lb of flour to bake a vanilla cake. She used another $\frac{3}{4}$ lb of flour to bake a chocolate cake. How much flour did she use altogether?

EUREKA MATH™

3. Carlos wants to practice piano 2 hours each day. He practices piano for $\frac{3}{4}$ hour before school and $\frac{7}{10}$ hour when he gets home. How many hours has Carlos practiced piano? How much longer does he need to practice before going to bed in order to meet his goal?

Name ______________________________ Date ________________

1. Draw a model to help solve $\frac{5}{6} + \frac{1}{4}$. Write your answer as a mixed number.

2. Patrick drank $\frac{3}{4}$ liter of water Monday before jogging. He drank $\frac{4}{5}$ liter of water after his jog. How much water did Patrick drink altogether? Write your answer as a mixed number.

Name ______________________________ Date ________________

1. For the following problems, draw a picture using the rectangular fraction model and write the answer. When possible, write your answer as a mixed number.

a. $\frac{3}{4} + \frac{1}{3} =$

b. $\frac{3}{4} + \frac{2}{3} =$

c. $\frac{1}{3} + \frac{3}{5} =$

d. $\frac{5}{6} + \frac{1}{2} =$

e. $\frac{2}{3} + \frac{5}{6} =$

f. $\frac{4}{3} + \frac{4}{7} =$

Solve the following problems. Draw a picture, and write the number sentence that proves the answer. Simplify your answer, if possible.

2. Sam made $\frac{2}{3}$ liter of punch and $\frac{3}{4}$ liter of tea to take to a party. How many liters of beverages did Sam bring to the party?

3. Mr. Sinofsky used $\frac{5}{8}$ of a tank of gas on a trip to visit relatives for the weekend and another 1 half of a tank commuting to work the next week. He then took another weekend trip and used $\frac{1}{4}$ tank of gas. How many tanks of gas did Mr. Sinofsky use altogether?

Lesson 5

Objective: Subtract fractions with unlike units using the strategy of creating equivalent fractions.

Suggested Lesson Structure

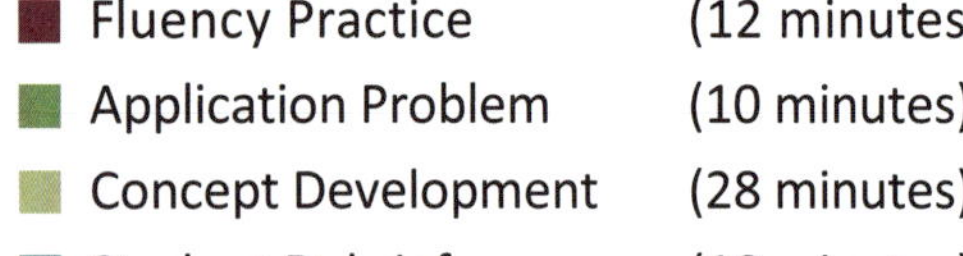

Fluency Practice	(12 minutes)
Application Problem	(10 minutes)
Concept Development	(28 minutes)
Student Debrief	(10 minutes)
Total Time	**(60 minutes)**

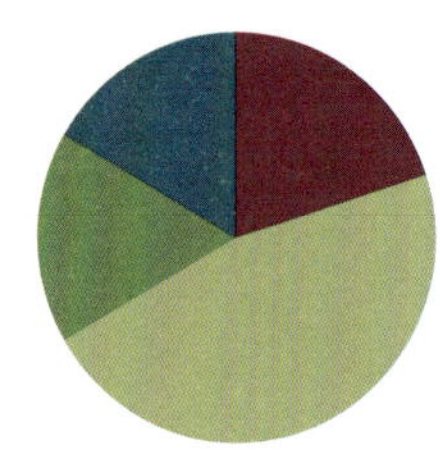

Fluency Practice (12 minutes)

- Sprint: Subtracting Fractions from a Whole Number **4.NF.3a** (12 minutes)

Sprint: Subtracting Fractions from a Whole Number (12 minutes)

Materials: (S) Subtracting Fractions from a Whole Number Sprint

Note: This Sprint is a quick mental exercise of part–part–whole understanding as it relates to fractions. (Between correcting Sprint A and giving Sprint B, have students share their strategies for quickly solving the problems. This very brief discussion may help some students catch on to a more efficient approach for Sprint B.)

Application Problem (10 minutes)

A farmer uses $\frac{3}{4}$ of his field to plant corn, $\frac{1}{6}$ of his field to plant beans, and the rest to plant wheat. What fraction of his field is used for wheat?

At times, simply remind students of their RDW process in order to solve a problem independently. It is desired that students internalize the simple set of questions as well as the systematic approach of read, draw, write an equation, and write a statement:

- What do I see?
- What can I draw?
- What conclusions can I make from my drawing?

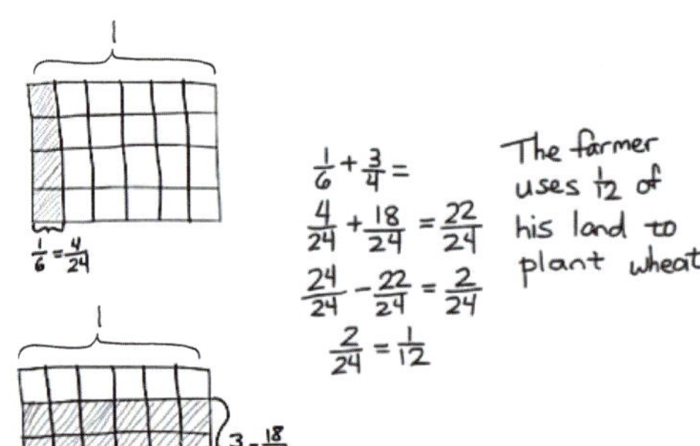

Note: Students solve this Application Problem involving addition and subtraction of fractions with unlike denominators, using visual models as learned in Lessons 3 and 4.

EUREKA MATH™

Concept Development (28 minutes)

Materials: (S) Personal white board

T: (Write 3 boys – 1 girl = ____.) Turn and talk to your partner about the answer.

S: You can't subtract 1 girl from 3 boys. You don't have any girls. → The answer is 2 students if you rename them as students. → The units are not the same, but we can rename them as students.

T: Yes. 3 students – 1 student = 2 students. (Write 1 half – 1 third.) What about 1 half minus 1 third? How is this problem the same as the one before? Turn and talk.

S: The units are not the same. → We have to change the units to find the difference.

NOTES ON MULTIPLE MEANS OF ENGAGEMENT:

If this problem is acted out, it can clarify confusion about units. Students will see that the group can be renamed *students* to encompass everyone and have like units.

Repeat the process with Problem 1 using pattern blocks. If the hexagon is the whole, the yellow trapezoid is $\frac{1}{2}$, the blue rhombus is $\frac{1}{3}$, and the green triangle is $\frac{1}{6}$.

Problem 1: $\frac{1}{2} - \frac{1}{3}$

T: (Write $\frac{1}{2} - \frac{1}{3}$.) We'll need to change both units.

T: I'll draw one fraction model and partition it into 2 equal units. Then I'll write 1 half below one part and shade it to make it easier to see what 1 half is after I change the units. (Model.)

T: On the second fraction model, I'll make thirds with horizontal lines and write 1 third next to it after shading it. (Model.)

T: Now, let's make equivalent units. (Model.) How many new units do we have?

S: 6 units.

T: 1 half is how many sixths?

S: 1 half is 3 sixths.

T: 1 third is how many sixths?

S: 1 third is 2 sixths.

T: (Write $\frac{1}{2} - \frac{1}{3} = \frac{3}{6} - \frac{2}{6}$. Cross out 2 of 3 shaded sixths.) Say the subtraction sentence with like units.

S: 3 sixths – 2 sixths = 1 sixth.

T: With unlike units?

S: 1 half – 1 third = 1 sixth.

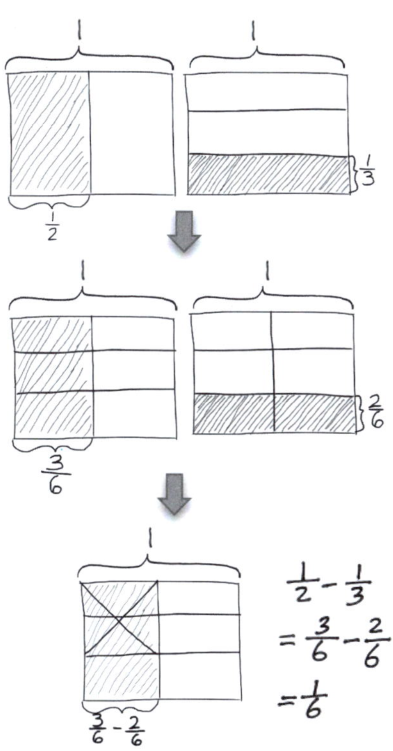

Problem 2: **a.** $\frac{1}{3} - \frac{1}{4}$ **b.** $\frac{1}{2} - \frac{1}{5}$

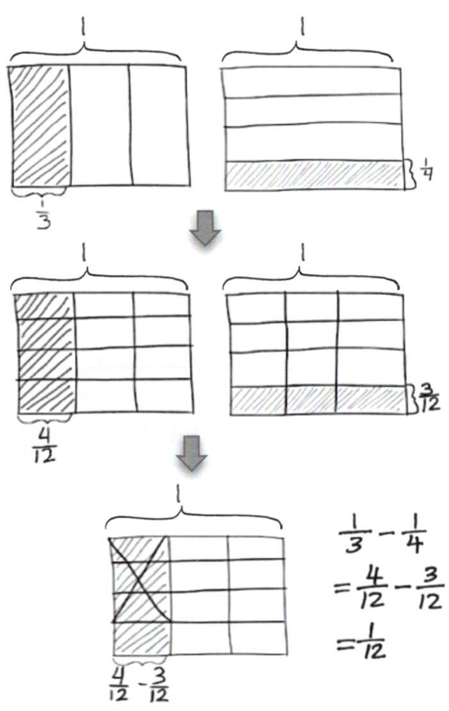

This next set of problems presents the additional complexity of partitioning a greater number of units.

T: (Write $\frac{1}{3} - \frac{1}{4}$.) Find the difference. Then, explain to your partner your strategy for solving.

S: To create like units, we can do exactly as we did when adding. We have to make smaller units. → First, we draw parts in one direction. Then, we partition in the other direction to find like units. → The only thing we have to remember is that we are subtracting the units, not adding.

T: What is our new smaller unit or common denominator?

S: Twelfths.

T: 1 third is...?

S: 4 twelfths.

T: 1 fourth is...?

S: 3 twelfths.

T: (Write $\frac{1}{3} - \frac{1}{4} = \frac{4}{12} - \frac{3}{12}$. Cross out three of the four twelfths.)

T: Say the subtraction sentence with like units.

S: 4 twelfths – 3 twelfths = 1 twelfth.

T: With unlike units?

S: 1 third – 1 fourth = 1 twelfth.

Repeat the process with the following suggested problem: $\frac{1}{2} - \frac{1}{5}$.

T: (Write $\frac{1}{2} - \frac{1}{5}$.) Solve this problem with a partner.

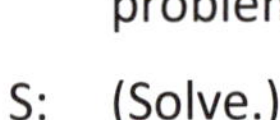

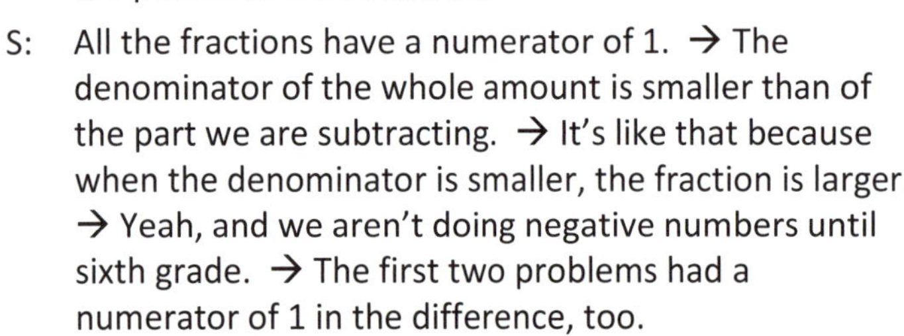

S: (Solve.)

T: What do you notice about all the problems we've solved?

S: All the fractions have a numerator of 1. → The denominator of the whole amount is smaller than of the part we are subtracting. → It's like that because when the denominator is smaller, the fraction is larger. → Yeah, and we aren't doing negative numbers until sixth grade. → The first two problems had a numerator of 1 in the difference, too.

T: I chose those problems for exactly that reason. Fractions with a numerator of 1 are called unit fractions. Let's try this next problem subtracting from a non-unit fraction.

NOTES ON MULTIPLE MEANS OF ENGAGEMENT:

Offering additional problems such as Problem 2 will allow students to obtain more practice if needed. If students are working above grade level, then prepare additional problems that challenge, but stay within the level standards.
For example, make a list of problems subtracting consecutive denominators.

$\frac{1}{5} - \frac{1}{6}$

$\frac{1}{6} - \frac{1}{7}$

$\frac{1}{7} - \frac{1}{8}$

Students working above grade level can look for patterns. Ask, "What pattern do you notice?"

Problem 3: $\frac{2}{3} - \frac{1}{4}$

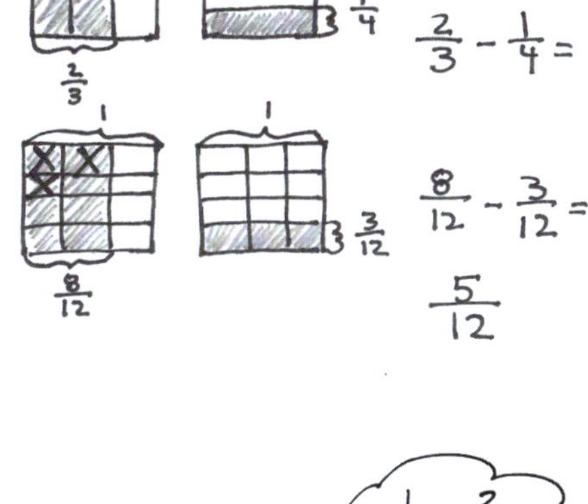

T: (Write $\frac{2}{3} - \frac{1}{4}$.) Discuss with your partner how you would solve this problem. Explain the difference in solving a problem when there is a non-unit fraction such as $\frac{2}{3}$ rather than $\frac{1}{3}$.

S: (Discuss.)

T: Work with a partner to solve.

S: (Solve.)

Problem 4: $\frac{1}{2} - \frac{2}{7}$

T: (Write $\frac{1}{2} - \frac{2}{7}$.) What is different about this problem?

S: It has a non-unit fraction being subtracted.

T: Very observant. Be careful when subtracting so that you take away the correct amount of units. Solve this problem with your partner.

S: (Solve.)

Problem 5: $\frac{4}{5} - \frac{2}{3}$

Here, students encounter both a whole and subtracted part, which are non-unit fractions.

T: (Write $\frac{4}{5} - \frac{2}{3}$.) Solve this problem.

S: (Solve.)

T: Turn and tell your partner how you labeled your rectangular fraction model. Compare your labeling of non-unit fractions with your labeling of unit fractions.

S: We have to label two rows if we want to show $\frac{2}{3}$. → Nothing really changes; we just bracket more parts.

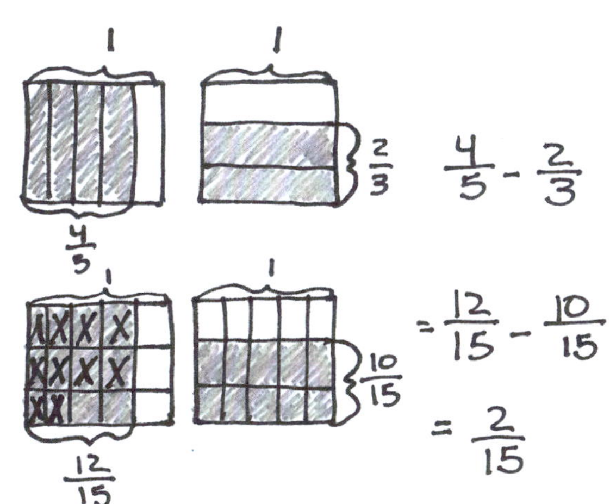

Problem Set (10 minutes)

Students should do their personal best to complete the Problem Set within the allotted 10 minutes. For some classes, it may be appropriate to modify the assignment by specifying which problems they work on first. Some problems do not specify a method for solving. Students should solve these problems using the RDW approach used for Application Problems.

Student Debrief (10 minutes)

Lesson Objective: Subtract fractions with unlike units using the strategy of creating equivalent fractions.

The Student Debrief is intended to invite reflection and active processing of the total lesson experience.

Invite students to review their solutions for the Problem Set. They should check work by comparing answers with a partner before going over answers as a class. Look for misconceptions or misunderstandings that can be addressed in the Debrief. Guide students in a conversation to debrief the Problem Set and process the lesson.

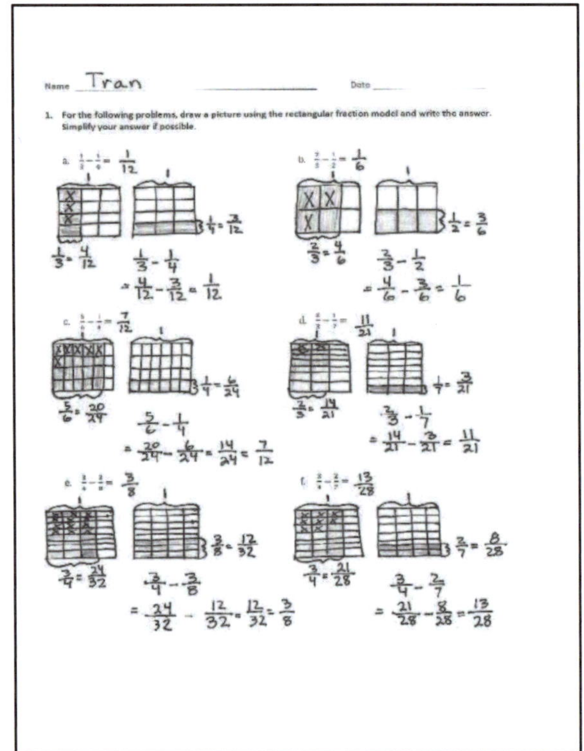

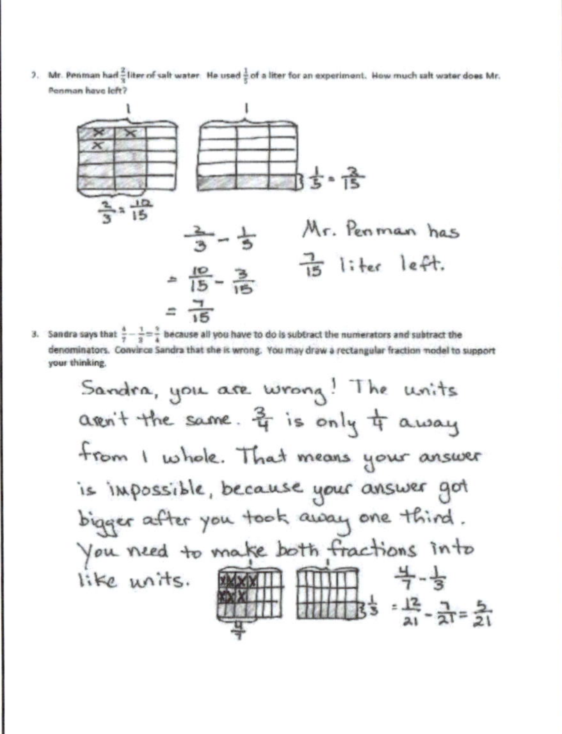

T: Bring your Problem Set to the Debrief. Take one minute to check your answers on Problems 1 and 2 with your partner. Do not change your answers, however. If you have a different answer, try to figure out why.

S: (Work.)

T: (Circulate. Look for common errors to guide your questioning during the next phase of the Student Debrief.)

T: I'll read the answers to Problems 1 and 2 now. (Read answers aloud.)

T: Review and correct your mistakes for two minutes. If you had no errors, please raise your hand. I will assign you to support a peer.

T: Compare with your partner. How do these problems relate to each other?

- 1(a) and 1(b)
- 1(b) and 1(d)
- 1(e) and 1(f)

Suggestions for facilitating the Student Debrief are as follows:

- Have students write about one relationship in their math journal.
- Have students do a pair–share.
- Meet with a small group of English language learners or students working below grade level while others do one of the above.
- Debrief the whole class after partner sharing.

- Circulate, and ask the following questions.
- Post the questions, and have student leaders facilitate small group discussions.

T: What do you notice about Problem 1 (a) and (b)?

S: $\frac{2}{3}$ is double $\frac{1}{3}$. → $\frac{1}{2}$ is double $\frac{1}{4}$, and $\frac{1}{6}$ is double $\frac{1}{12}$.

T: What do you notice about Problem 1 (b) and (d)?

S: Both problems start with $\frac{2}{3}$. → $\frac{2}{3}$ is the whole in both, but in one problem, you are taking away $\frac{1}{2}$ renamed as 3 sixths. → When you are subtracting $\frac{3}{21}$, you are taking away 3 much smaller units. → That means the answer to Problem 1(b) is greater. → $\frac{1}{6}$ is less than $\frac{11}{21}$. → Yeah, $\frac{11}{21}$ is a little more than a half. Half of 21 is 10.5. Eleven is greater than that. → $\frac{1}{6}$ is closer to zero.

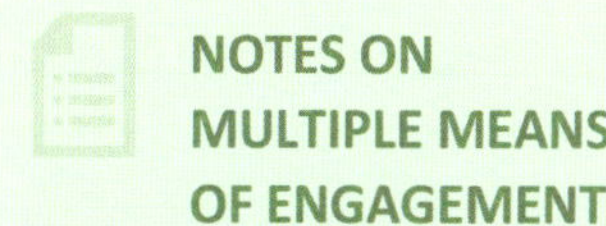

NOTES ON MULTIPLE MEANS OF ENGAGEMENT:

Meet with a small group while the rest of the students complete the Debrief activities independently.

T: What do you notice about Problem 1 (e) and (f)?

S: Both problems start with $\frac{3}{4}$. But in one, you are taking away $\frac{3}{8}$, and in the other, you are taking away $\frac{2}{7}$. → $\frac{3}{8}$ is half of $\frac{3}{4}$. → Yeah, $\frac{3}{8}$ doubled is $\frac{3}{4}$. → $\frac{13}{28}$ is $\frac{1}{28}$ away from a half, but $\frac{3}{8}$ is $\frac{1}{8}$ less than a half. $\frac{13}{28}$ is a greater answer, so $\frac{2}{7}$ must be less than $\frac{3}{8}$.

T: Share the strategies you used to solve the word problems.

S: (Share.)

T: If you were going to design a Problem Set for this lesson, what would you have done differently? Would you have included as many unit fractions? More word problems?

S: (Share.)

Exit Ticket (3 minutes)

After the Student Debrief, instruct students to complete the Exit Ticket. A review of their work will help with assessing students' understanding of the concepts that were presented in today's lesson and planning more effectively for future lessons. The questions may be read aloud to the students.

A

Number Correct: _______

Subtracting Fractions from a Whole Number

1.	$4 - \frac{1}{2} =$			23.	$3 - \frac{1}{8} =$	
2.	$3 - \frac{1}{2} =$			24.	$3 - \frac{3}{8} =$	
3.	$2 - \frac{1}{2} =$			25.	$3 - \frac{5}{8} =$	
4.	$1 - \frac{1}{2} =$			26.	$3 - \frac{7}{8} =$	
5.	$1 - \frac{1}{3} =$			27.	$2 - \frac{7}{8} =$	
6.	$2 - \frac{1}{3} =$			28.	$4 - \frac{1}{7} =$	
7.	$4 - \frac{1}{3} =$			29.	$3 - \frac{6}{7} =$	
8.	$4 - \frac{2}{3} =$			30.	$2 - \frac{3}{7} =$	
9.	$2 - \frac{2}{3} =$			31.	$4 - \frac{4}{7} =$	
10.	$2 - \frac{1}{4} =$			32.	$3 - \frac{5}{7} =$	
11.	$2 - \frac{3}{4} =$			33.	$4 - \frac{3}{4} =$	
12.	$3 - \frac{3}{4} =$			34.	$2 - \frac{5}{8} =$	
13.	$3 - \frac{1}{4} =$			35.	$3 - \frac{3}{10} =$	
14.	$4 - \frac{3}{4} =$			36.	$4 - \frac{2}{5} =$	
15.	$2 - \frac{1}{10} =$			37.	$4 - \frac{3}{7} =$	
16.	$3 - \frac{9}{10} =$			38.	$3 - \frac{7}{10} =$	
17.	$2 - \frac{7}{10} =$			39.	$3 - \frac{5}{10} =$	
18.	$4 - \frac{3}{10} =$			40.	$4 - \frac{2}{8} =$	
19.	$3 - \frac{1}{5} =$			41.	$2 - \frac{9}{12} =$	
20.	$3 - \frac{2}{5} =$			42.	$4 - \frac{2}{12} =$	
21.	$3 - \frac{4}{5} =$			43.	$3 - \frac{2}{6} =$	
22.	$3 - \frac{3}{5} =$			44.	$2 - \frac{8}{12} =$	

B

Number Correct: ______

Improvement: ______

Subtracting Fractions from a Whole Number

1.	$1 - \frac{1}{2} =$	
2.	$2 - \frac{1}{2} =$	
3.	$3 - \frac{1}{2} =$	
4.	$4 - \frac{1}{2} =$	
5.	$1 - \frac{1}{4} =$	
6.	$2 - \frac{1}{4} =$	
7.	$4 - \frac{1}{4} =$	
8.	$4 - \frac{3}{4} =$	
9.	$2 - \frac{3}{4} =$	
10.	$2 - \frac{1}{3} =$	
11.	$2 - \frac{2}{3} =$	
12.	$3 - \frac{2}{3} =$	
13.	$3 - \frac{1}{3} =$	
14.	$4 - \frac{2}{3} =$	
15.	$3 - \frac{1}{10} =$	
16.	$2 - \frac{9}{10} =$	
17.	$4 - \frac{7}{10} =$	
18.	$3 - \frac{3}{10} =$	
19.	$2 - \frac{1}{5} =$	
20.	$2 - \frac{2}{5} =$	
21.	$2 - \frac{4}{5} =$	
22.	$3 - \frac{3}{5} =$	

23.	$2 - \frac{1}{8} =$	
24.	$2 - \frac{3}{8} =$	
25.	$2 - \frac{5}{8} =$	
26.	$2 - \frac{7}{8} =$	
27.	$4 - \frac{7}{8} =$	
28.	$3 - \frac{1}{7} =$	
29.	$2 - \frac{6}{7} =$	
30.	$4 - \frac{3}{7} =$	
31.	$3 - \frac{4}{7} =$	
32.	$2 - \frac{5}{7} =$	
33.	$3 - \frac{3}{4} =$	
34.	$4 - \frac{5}{8} =$	
35.	$2 - \frac{3}{10} =$	
36.	$3 - \frac{2}{5} =$	
37.	$3 - \frac{3}{7} =$	
38.	$2 - \frac{7}{10} =$	
39.	$2 - \frac{5}{10} =$	
40.	$3 - \frac{6}{8} =$	
41.	$4 - \frac{3}{12} =$	
42.	$3 - \frac{10}{12} =$	
43.	$2 - \frac{4}{6} =$	
44.	$4 - \frac{4}{12} =$	

Name ______________________________ Date ______________

1. For the following problems, draw a picture using the rectangular fraction model and write the answer. Simplify your answer, if possible.

a. $\frac{1}{3} - \frac{1}{4} =$

b. $\frac{2}{3} - \frac{1}{2} =$

c. $\frac{5}{6} - \frac{1}{4} =$

d. $\frac{2}{3} - \frac{1}{7} =$

e. $\frac{3}{4} - \frac{3}{8} =$

f. $\frac{3}{4} - \frac{2}{7} =$

2. Mr. Penman had $\frac{2}{3}$ liter of salt water. He used $\frac{1}{5}$ of a liter for an experiment. How much salt water does Mr. Penman have left?

3. Sandra says that $\frac{4}{7} - \frac{1}{3} = \frac{3}{4}$ because all you have to do is subtract the numerators and subtract the denominators. Convince Sandra that she is wrong. You may draw a rectangular fraction model to support your thinking.

Name ____ Date ____________________

For the following problems, draw a picture using the rectangular fraction model and write the answer. Simplify your answer, if possible.

a. $\frac{1}{2} - \frac{1}{7} =$

b. $\frac{3}{5} - \frac{1}{2} =$

Name ______________________________ Date ____________

1. The picture below shows $\frac{3}{4}$ of the rectangle shaded. Use the picture to show how to create an equivalent fraction for $\frac{3}{4}$, and then subtract $\frac{1}{3}$.

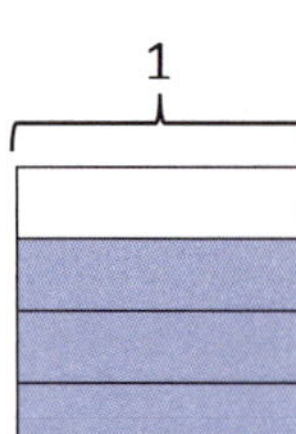

$\frac{3}{4} - \frac{1}{3} =$

2. Find the difference. Use a rectangular fraction model to find common denominators. Simplify your answer, if possible.

a. $\frac{5}{6} - \frac{1}{3} =$

b. $\frac{2}{3} - \frac{1}{2} =$

c. $\frac{5}{6} - \frac{1}{4} =$

d. $\frac{4}{5} - \frac{1}{2} =$

e. $\frac{2}{3} - \frac{2}{5} =$

f. $\frac{5}{7} - \frac{2}{3} =$

3. Robin used $\frac{1}{4}$ of a pound of butter to make a cake. Before she started, she had $\frac{7}{8}$ of a pound of butter. How much butter did Robin have when she was done baking? Give your answer as a fraction of a pound.

4. Katrina needs $\frac{3}{5}$ kilogram of flour for a recipe. Her mother has $\frac{3}{7}$ kilogram of flour in her pantry. Is this enough flour for the recipe? If not, how much more will she need?

Lesson 6

Objective: Subtract fractions from numbers between 1 and 2.

Suggested Lesson Structure

Fluency Practice	(10 minutes)
Application Problem	(8 minutes)
Concept Development	(32 minutes)
Student Debrief	(10 minutes)
Total Time	**(60 minutes)**

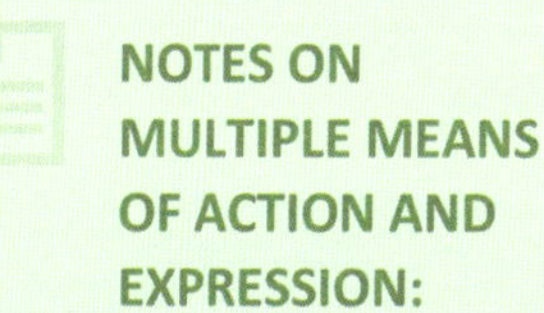

Fluency Practice (10 minutes)

- Name the Fraction to Complete the Whole **4.NF.3b** (4 minutes)
- Taking from the Whole **5.NF.7** (3 minutes)
- Fraction Units to Ones and Fractions **5.NF.7** (3 minutes)

Name the Fraction to Complete the Whole (4 minutes)

Note: This fluency activity is a quick mental exercise of part–part–whole understanding as it relates to fractions.

T: I'll say a fraction, and you say the missing part to make one whole. Ready? $\frac{1}{2}$.

S: $\frac{1}{2}$.

T: $\frac{4}{5}$.

S: $\frac{1}{5}$.

T: $\frac{1}{7}$.

S: $\frac{6}{7}$.

T: $\frac{4}{9}$.

S: $\frac{5}{9}$.

T: $\frac{18}{20}$.

S: $\frac{2}{20}$.

NOTES ON MULTIPLE MEANS OF ACTION AND EXPRESSION:

As a variation to the Name the Fraction to Complete the Whole fluency activity, have students quiz each other. Homogeneous groups may be beneficial.

For students working below grade level, provide a bar diagram template in their personal white boards so that students can quickly draw each fraction and see the unknown or missing part.

For students working above grade level, give them $\frac{1}{2}$ as a target number. Their partner can give them any fraction less than one. They tell how much to add or subtract to arrive at one half. For example, $\frac{3}{7}$ → Add $\frac{1}{14}$; $\frac{9}{10}$ → Subtract $\frac{4}{10}$.

T: $\frac{147}{150}$.

S: $\frac{3}{150}$.

T: Share your strategy for making one whole with a partner.

T: With your partner, take turns giving each other problems to solve. You have one minute.

Taking from the Whole (3 minutes)

Materials: (S) Personal white board

Note: This fluency activity strengthens mental math and lays the foundation for today's Concept Development in which students subtract from numbers between 1 and 2.

T: I'll say a subtraction expression. You say the answer. 1 – 1 half.

S: 1 half.

T: 1 – 1 third.

S: 2 thirds.

T: 1 – 2 thirds.

S: 1 third.

T: 1 – 2 fifths.

S: 3 fifths.

T: 1 – 4 fifths.

S: 1 fifth.

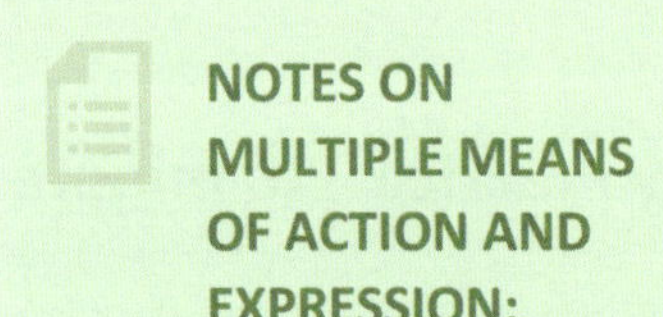

NOTES ON MULTIPLE MEANS OF ACTION AND EXPRESSION:

If students struggle to answer chorally, write the subtraction sentences in numerical form on the board. Have students answer the problems on their personal white boards.

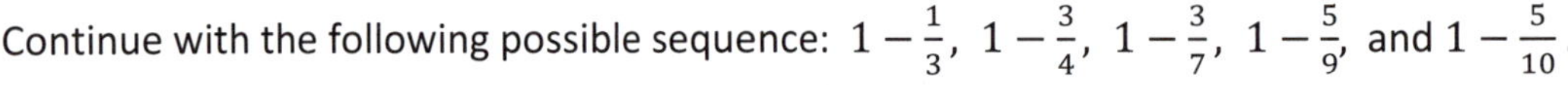

Continue with the following possible sequence: $1-\frac{1}{3}$, $1-\frac{3}{4}$, $1-\frac{3}{7}$, $1-\frac{5}{9}$, and $1-\frac{5}{10}$.

Fraction Units to Ones and Fractions (3 minutes)

Note: Students rapidly and mentally generate mixed numbers that are equivalent to fractions greater than 1 in preparation for today's Concept Development.

T: I'll say a fraction; you say it as a mixed number. Three halves.

S: One and one half.

T: Five halves.

S: Two and one half.

T: Seven halves.

S: Three and one half.

T: Eleven halves.

S: Five and one half.

Continue with the following possible sequence: $\frac{4}{3}$, $\frac{5}{3}$, $\frac{10}{3}$, $\frac{22}{3}$, and $\frac{5}{4}$, $\frac{7}{4}$, $\frac{11}{4}$, $\frac{39}{4}$.

Application Problem (8 minutes)

The Napoli family combined two bags of dry cat food in a plastic container. One bag had $\frac{5}{6}$ kg of cat food. The other bag had $\frac{3}{4}$ kg. What was the total weight of the container after the bags were combined?

T: Use the RDW process to solve the problem independently. Use your questions to support you in your work. What do you see? Can you draw something? What conclusions can you make from your drawing?

T: We will analyze two solution strategies in four minutes.

After four minutes, lead students through a brief comparison of a more concrete strategy such as the one below, on the left, and the more abstract strategy below, on the right. Ensure students realize that both answers, $1\frac{7}{12}$ and $1\frac{14}{24}$, are correct.

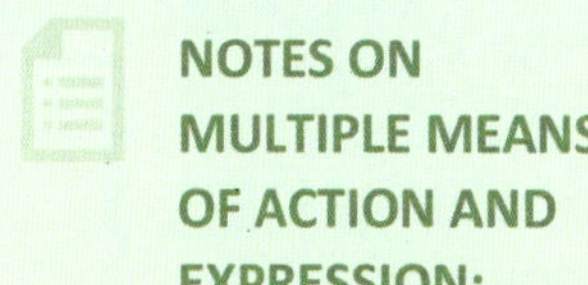

NOTES ON MULTIPLE MEANS OF ACTION AND EXPRESSION:

At this point, some students may realize they can combine their drawings onto one model, rather than drawing them separately as in previous lessons. Students working above grade level should be encouraged to combine their drawings into one model.

Solution 1

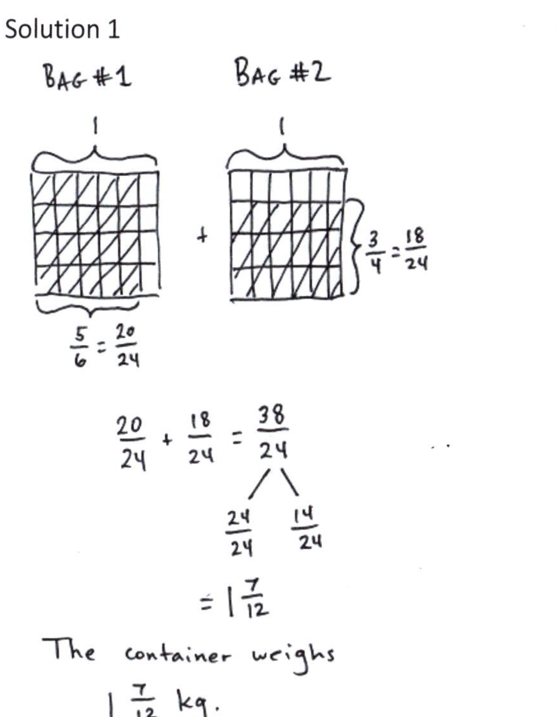

Solution 2

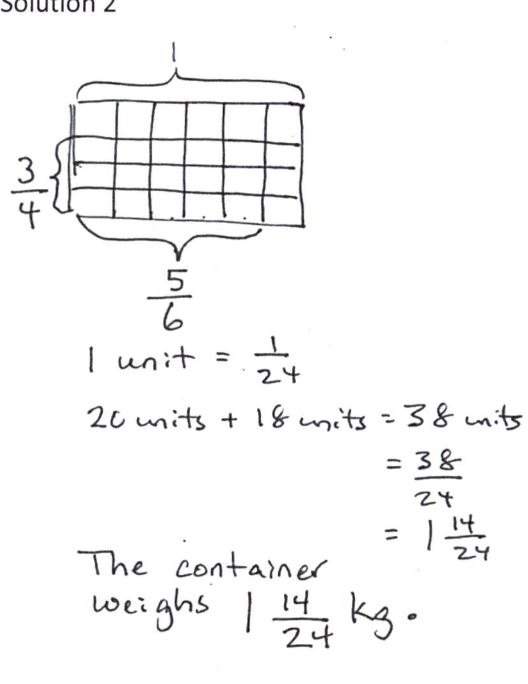

Note: This Application Problem reviews addition of fractions with unlike denominators, using visual models, and connects to today's subtraction of unlike units (between 1 and 2).

Concept Development (32 minutes)

Materials: (S) Personal white board

Problem 1: $1\frac{1}{3} - \frac{1}{2}$

Method 1 Method 2

$1\frac{1}{3}$

$1\frac{1}{3} - \frac{1}{2} = \frac{3}{6} + \frac{2}{6}$ $\frac{6}{6}$ $\frac{2}{6}$ $= \frac{5}{6}$

or

$1\frac{1}{3} - \frac{1}{2} = \frac{8}{6} - \frac{3}{6}$ $= \frac{5}{6}$

T: (Write $1\frac{1}{3} - \frac{1}{2}$.) Read the subtraction expression.

S: 1 and 1 third minus 1 half.

T: How many thirds is 1 and 1 third?

S: 4 thirds.

T: (Draw 1 whole and 1 third.) What should we do now? Turn and talk to your partner.

S: Make like units.

T: How many new smaller units are in each whole?

S: 6 units.

T: 4 thirds is how many sixths?

S: 8 sixths.

T: 1 half is how many sixths?

S: 3 sixths.

T: Looking at my drawing, how would you subtract 3 sixths or a half? Discuss this with your partner.

S: You can take the half from the whole and then add back the third. → Then, you are adding to subtract? → Yes, you are adding the part you had left after you take away. → It makes it easier because we know really well how to subtract any fraction from one whole. → Yeah, but it's just easier for me to take the 3 sixths from the 8 sixths. → For me, it's easier to take it from the whole and add back the rest.

T: It's like subtracting 80 from 130. It's easier for me to take 80 from 100 and add 20 and 30.

S: Can we subtract it either way?

T: Of course. Choose the way that is easiest for you.

T: Let's call the different solution strategies Method 1 and Method 2. If you use Method 1, let's record using a number bond. Solve and share your solution with a partner.

S: (Solve and share.) 8 sixths – 3 sixths = 5 sixths. → 1 and 1 third – 1 half = 5 sixths.

NOTES ON MULTIPLE MEANS OF ENGAGEMENT:

Have students use their personal white boards to follow along with the drawings that are demonstrated on the board so they can match the language with the model and the steps of the process. At key moments, have students orally label the parts of the model to practice using language.

Problem 2: $1\frac{1}{5} - \frac{1}{3}$

T: (Write $1\frac{1}{5} - \frac{1}{3}$.) I'll draw one rectangle to show 1 and a second rectangle to show 1 fifth. (Model.)

T: Are these units the same? Can I use fifths to subtract thirds?

S: No.

T: Explain to your partner how to solve this problem. Use both methods.

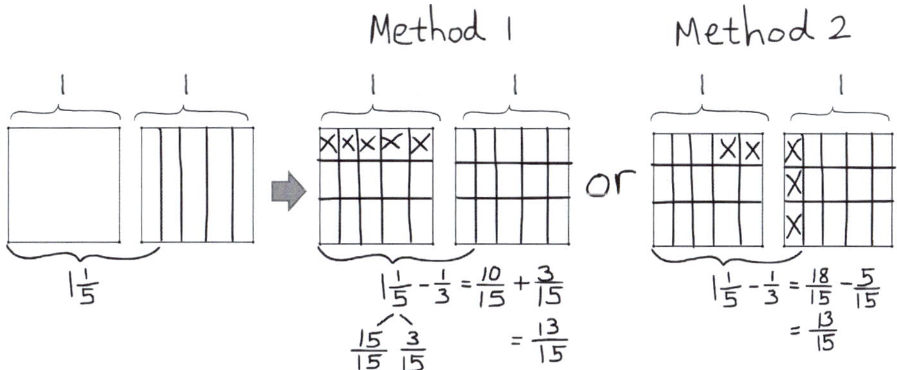

Problem 3: $1\frac{1}{2} - \frac{2}{3}$

The additional complexity here is the subtraction of a non-unit fraction.

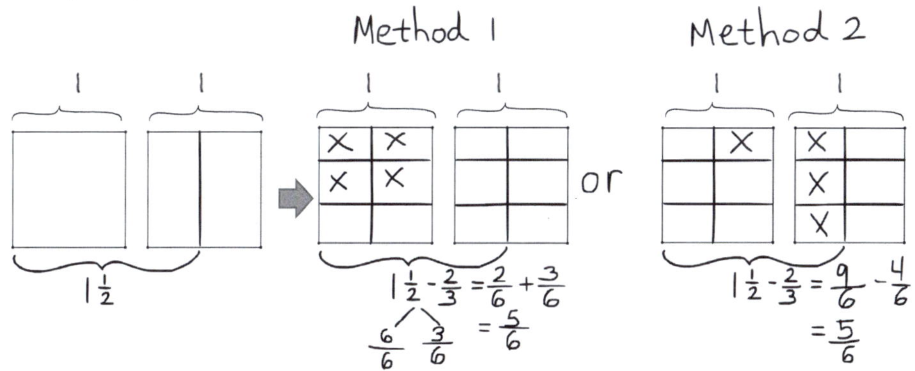

Problem 4: $1\frac{3}{4} - \frac{4}{5}$

In this problem, the new complexity is the use of two non-unit fractions.

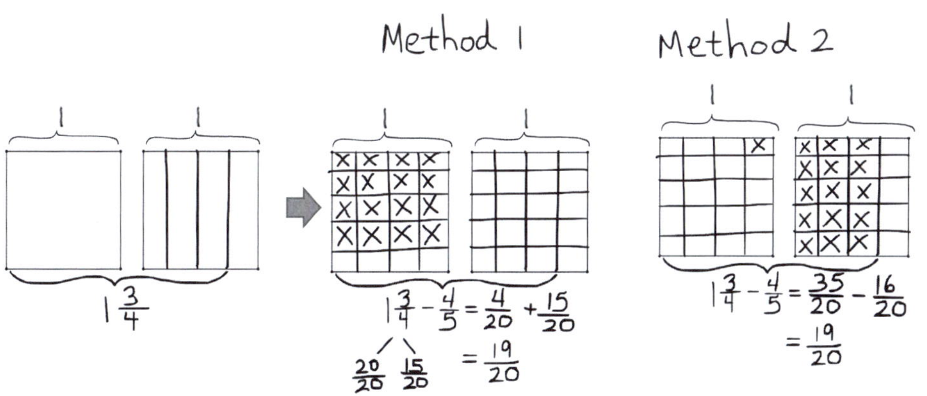

Problem 5: $1\frac{4}{9} - \frac{1}{2}$

T: (After students work, display Method 1 on the board.) Tell your neighbor what alternate strategy was used in this model.

S: (Share.)

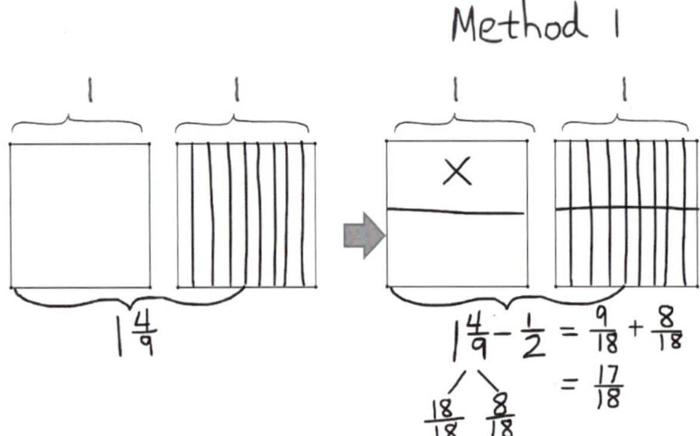

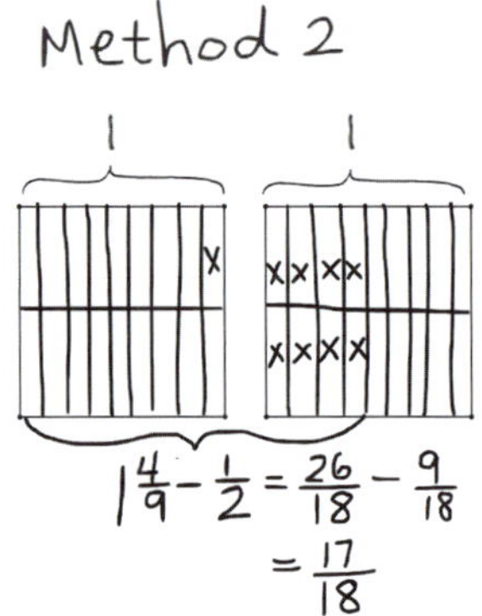

Problem Set (10 minutes)

Students should do their personal best to complete the Problem Set within the allotted 10 minutes. For some classes, it may be appropriate to modify the assignment by specifying which problems they work on first. Some problems do not specify a method for solving. Students should solve these problems using the RDW approach used for Application Problems.

NOTES ON MULTIPLE MEANS OF ACTION AND EXPRESSION:

As students share out various strategies, use a modified text representation activity. Have the rest of the class demonstrate on personal white boards the ideas their peers orally express.

Student Debrief (10 minutes)

Lesson Objective: Subtract fractions from numbers between 1 and 2.

The Student Debrief is intended to invite reflection and active processing of the total lesson experience.

Invite students to review their solutions for the Problem Set. They should check work by comparing answers with a partner before going over answers as a class. Look for misconceptions or misunderstandings that can be addressed in the Debrief. Guide students in a conversation to debrief the Problem Set and process the lesson.

T: Take one minute to compare your work with a partner's. (Circulate and look for common errors or student work to use instructionally.)

T: I'll read the answers to Problems 1 and 2 now. (Read answers aloud.)

T: (Students correct their work for about 2 minutes.) If you had no errors, I will assign you to support a peer.

T: Compare these problems with a partner:

- 1(a) and (b)
- 1(c) and (d)
- 1(e) and (f)

S: I remember that $\frac{1}{3} - \frac{1}{4}$ is $\frac{1}{12}$, so then $1\frac{1}{4} - \frac{1}{3}$ is $\frac{1}{12}$ less than 1. → It's the same with $1\frac{1}{3} - \frac{1}{5}$; the answer is $\frac{2}{15}$ less than 1. → You could use the same strategy on all of them.

T: Jacqueline, can you explain your solution to Problem 2?

S: I realized that the problem was really easy. It's just subtraction. I could take $\frac{5}{6}$ from 1 and add it to $\frac{1}{4}$. $\frac{1}{6}$ and $\frac{1}{4}$ are easy because they are just unit fractions $\frac{4}{24}$ and $\frac{6}{24}$. So, the answer is 10 twenty-fourths.

T: Did anyone solve it differently?

S: Yes. I just converted the fractions to like units and subtracted. So, it was 24 twenty-fourths and 6 twenty-fourths. 30 twenty-fourths – 20 twenty-fourths = 10 twenty-fourths.

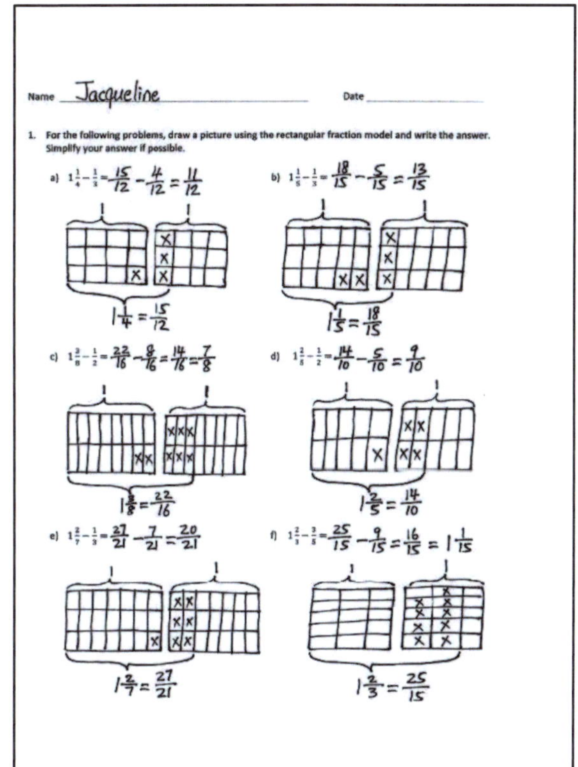

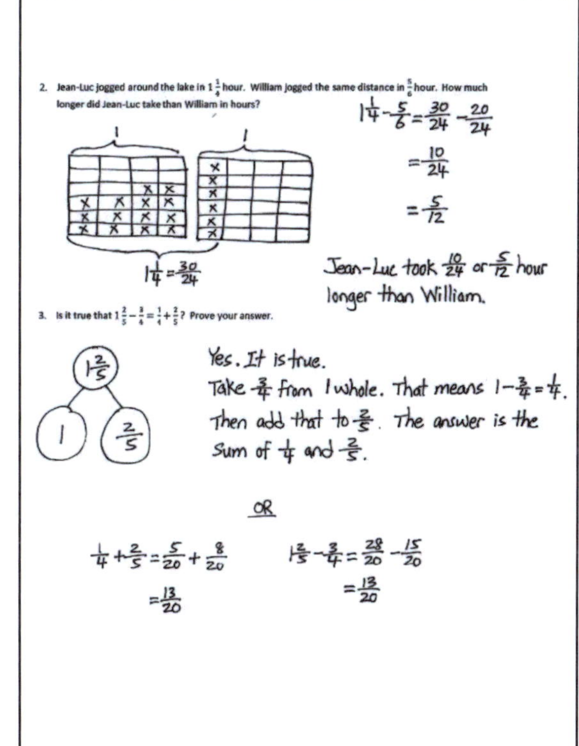

Exit Ticket (3 minutes)

After the Student Debrief, instruct students to complete the Exit Ticket. A review of their work will help with assessing students' understanding of the concepts that were presented in today's lesson and planning more effectively for future lessons. The questions may be read aloud to the students.

Name ______________________________ Date ______________

1. For the following problems, draw a picture using the rectangular fraction model and write the answer. Simplify your answer, if possible.

a. $1\frac{1}{4} - \frac{1}{3} =$

b. $1\frac{1}{5} - \frac{1}{3} =$

c. $1\frac{3}{8} - \frac{1}{2} =$

d. $1\frac{2}{5} - \frac{1}{2} =$

e. $1\frac{2}{7} - \frac{1}{3} =$

f. $1\frac{2}{3} - \frac{3}{5} =$

2. Jean-Luc jogged around the lake in $1\frac{1}{4}$ hour. William jogged the same distance in $\frac{5}{6}$ hour. How much longer did Jean-Luc take than William in hours?

3. Is it true that $1\frac{2}{5} - \frac{3}{4} = \frac{1}{4} + \frac{2}{5}$? Prove your answer.

Name ______________________________ Date ______________

For the following problems, draw a picture using the rectangular fraction model and write the answer. Simplify your answer, if possible.

a. $1\frac{1}{5} - \frac{1}{2} =$

b. $1\frac{1}{3} - \frac{5}{6} =$

Name ____ Date ____________________

1. For the following problems, draw a picture using the rectangular fraction model and write the answer. Simplify your answer, if possible.

a. $1 - \frac{5}{6} =$

b. $\frac{3}{2} - \frac{5}{6} =$

c. $\frac{4}{3} - \frac{5}{7} =$

d. $1\frac{1}{8} - \frac{3}{5} =$

e. $1\frac{2}{5} - \frac{3}{4} =$

f. $1\frac{5}{6} - \frac{7}{8} =$

g. $\frac{9}{7} - \frac{3}{4} =$

h. $1\frac{3}{12} - \frac{2}{3} =$

2. Sam had $1\frac{1}{2}$ m of rope. He cut off $\frac{5}{8}$ m and used it for a project. How much rope does Sam have left?

3. Jackson had $1\frac{3}{8}$ kg of fertilizer. He used some to fertilize a flower bed, and he only had $\frac{2}{3}$ kg left. How much fertilizer was used in the flower bed?

Lesson 7

Objective: Solve two-step word problems.

Suggested Lesson Structure

Fluency Practice	(12 minutes)
Concept Development	(38 minutes)
Student Debrief	(10 minutes)
Total Time	**(60 minutes)**

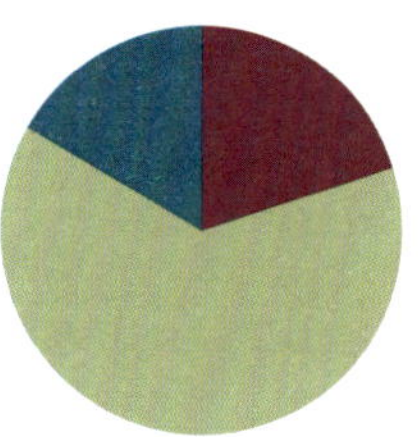

Fluency Practice (12 minutes)

- Sprint: Circle the Equivalent Fraction **4.NF.2** (12 minutes)

Sprint: Circle the Equivalent Fraction (12 minutes)

Materials: (S) Circle the Equivalent Fraction Sprint

Note: Students rapidly recognize common equivalent fractions mentally (i.e., without using multiplication or division).

Concept Development (38 minutes)

Materials: (S) Problem Set, personal white board

Note: For this lesson, the Problem Set comprises word problems from the Concept Development and is therefore to be used during the lesson itself.

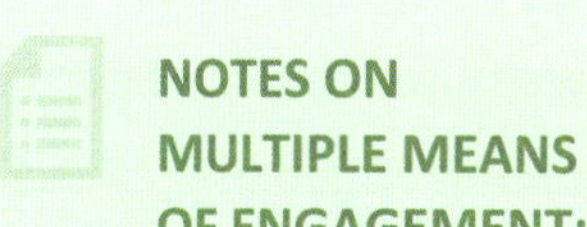

NOTES ON MULTIPLE MEANS OF ENGAGEMENT:

Consider strategically pairing English language learners and students working below grade level. For example, a student working below grade level who is accustomed to translating for a parent may blossom when asked to translate for a newcomer. The relationship may become mutually beneficial if the newcomer exhibits strong abilities and can help his partner with math concepts.

Problem 1

George weeded $\frac{1}{5}$ of the garden, and Summer weeded some, too. When they were finished, $\frac{2}{3}$ of the garden still needed to be weeded. What fraction of the garden did Summer weed?

T: Let's read the problem together.

S: (Read chorally.)

T: Share with your partner: What do you see when you hear the story? What can you draw?

S: (Share.)

T: I'll give you one minute to draw.

S: (Draw.)

T: What fraction of the garden did Summer weed? Is it a part or the whole?

S: Part.

T: Do we know the whole?

S: Yes.

T: What is it?

S: 1.

T: From the whole, we separate $\frac{1}{5}$ for George, an unknown amount for Summer, and have a leftover part of $\frac{2}{3}$. How do you solve for Summer's part? Turn and share.

S: (Share.)

T: Solve the problem on your personal white board. (Pause.) Show your board.

T: Turn and explain to your partner how you got the answer.

S: (Share.)

T: Jason, please share.

S: After I drew the tape diagram, I just subtracted the part George weeded and the part that was left from the whole. (See Solution 1.)

T: Barbara, please share.

S: My way to solve this problem is to add up the 2 parts to create a larger part and then subtract from the whole. (See Solution 2.)

T: What fraction of the garden did Summer weed?

S: Summer weeded $\frac{2}{15}$ of the garden.

T: Barbara and Jason have presented their solution strategies, which came directly from their drawings. With your partner, analyze their drawings. How are they the same, and how are they different?

S: (Compare and discuss.)

T: Are they both correct?

S: Yes.

T: How do you know?

S: They each make sense. → They each got the correct answer. → They each showed the same relationships but in different ways.

Solution 1

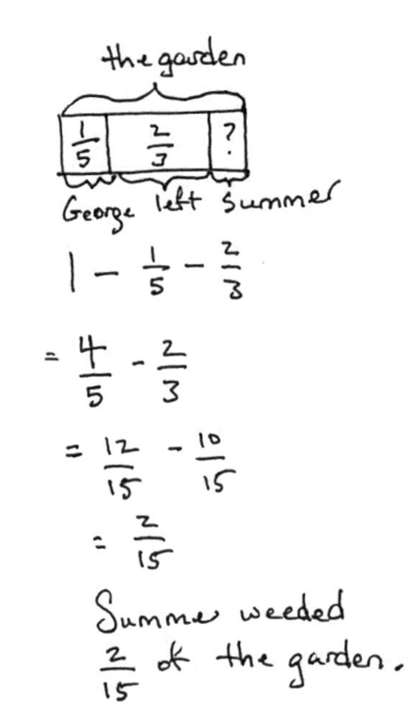

Solution 2

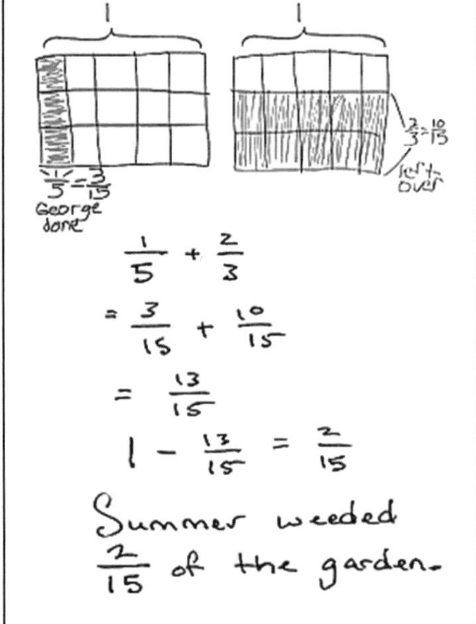

Problem 2

Jing spent $\frac{1}{3}$ of her money on a pack of pens, $\frac{1}{2}$ of her money on a pack of markers, and $\frac{1}{8}$ of her money on a pack of pencils. What fraction of her money is left?

Solution 1

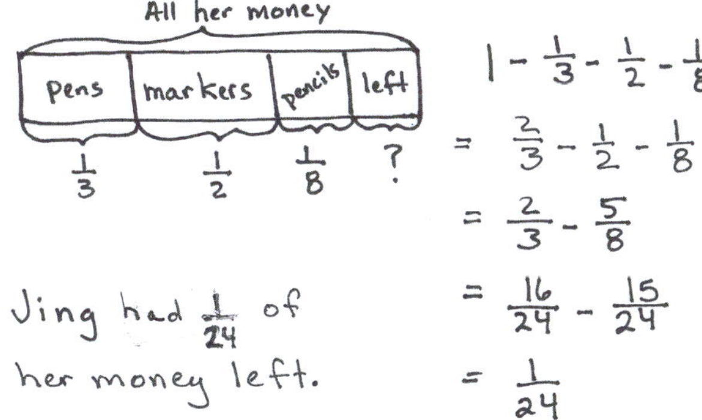

Solution 2

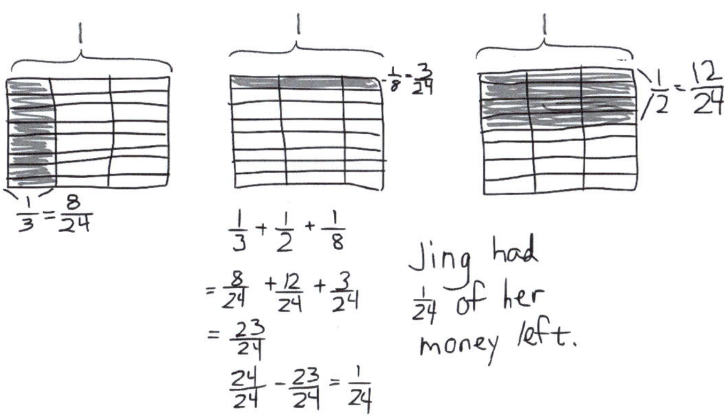

NOTES ON MULTIPLE MEANS OF ACTION AND EXPRESSION:

If students finish quickly, have them rework problems using a different strategy. Then, during the Student Debrief, take a moment to have them share which strategy they prefer and why.

Some students may realize they can find the number of like units using one rectangular fraction model. This is more efficient, and many students might benefit from this shortcut.

In the Student Debrief of the problem, compare the methods and support the validity of this strategy.

Problem 3

Shelby bought a 2-ounce tube of blue paint. She used $\frac{2}{3}$ ounce to paint the water, $\frac{3}{5}$ ounce to paint the sky, and some to paint a flag. After that, she had $\frac{2}{15}$ ounce left. How much paint did Shelby use to paint her flag?

Solution 1

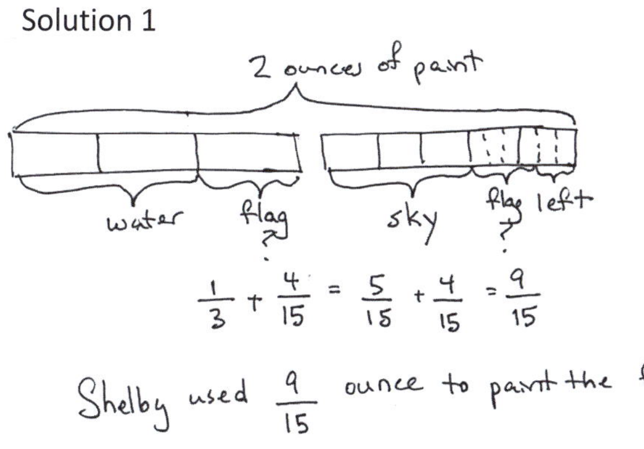

Solution 2

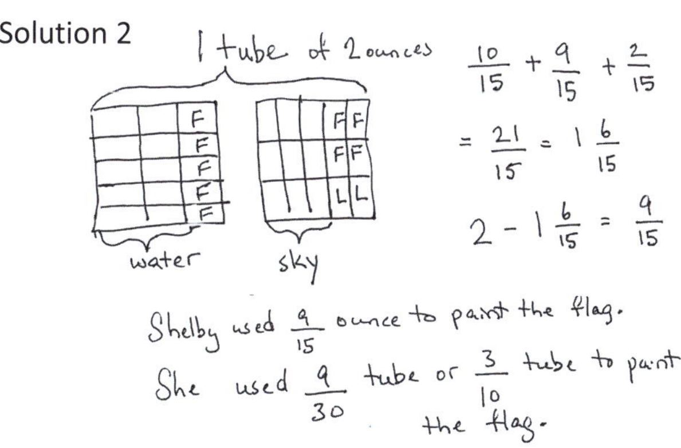

Problem 4

Jim sold $\frac{3}{4}$ gallon of lemonade. Dwight sold some lemonade, too. Together, they sold $1\frac{5}{12}$ gallons. Who sold more lemonade, Jim or Dwight? How much more? (See the Student Debrief for student work samples.)

Solution 1

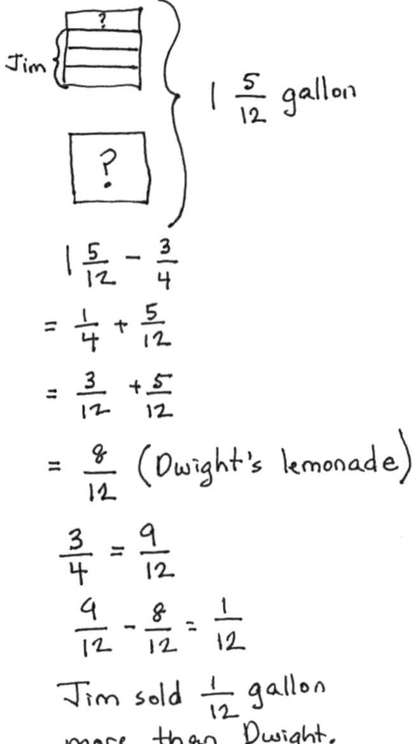

Solution 2

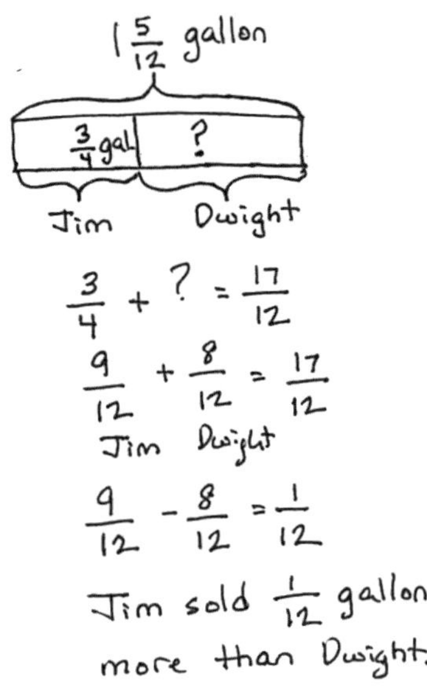

Problem 5

Leonard spent $\frac{1}{4}$ of his money on a sandwich. He spent 2 times as much on a gift for his brother as on some comic books. He had $\frac{3}{8}$ of his money left. What fraction of his money did he spend on the comic books?

Solution 1

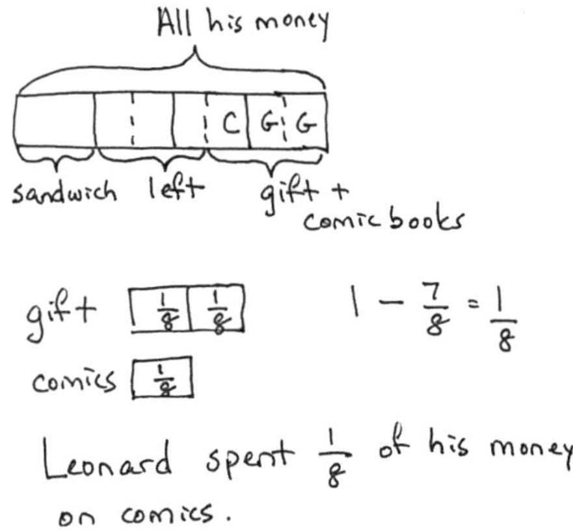

Solution 2

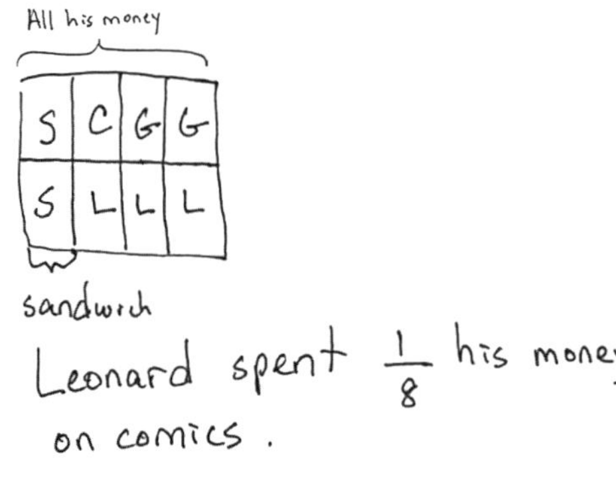

Student Debrief (10 minutes)

Lesson Objective: Solve two-step word problems.

The Student Debrief is intended to invite reflection and active processing of the total lesson experience.

Invite students to review their solutions for the Problem Set. They should check work by comparing answers with a partner before going over answers as a class. Look for misconceptions or misunderstandings that can be addressed in the Debrief. Guide students in a conversation to debrief the Problem Set and process the lesson.

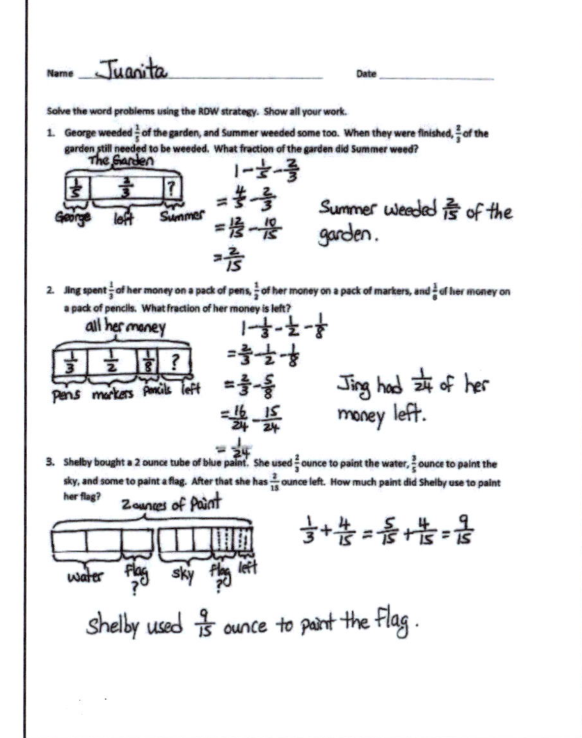
Name Juanita Date

Solve the word problems using the RDW strategy. Show all your work.

1. George weeded $\frac{1}{5}$ of the garden, and Summer weeded some too. When they were finished, $\frac{2}{3}$ of the garden still needed to be weeded. What fraction of the garden did Summer weed?

$1 - \frac{1}{5} - \frac{2}{3}$
$= \frac{4}{5} - \frac{2}{3}$
$= \frac{12}{15} - \frac{10}{15}$
$= \frac{2}{15}$

Summer weeded $\frac{2}{15}$ of the garden.

2. Jing spent $\frac{1}{3}$ of her money on a pack of pens, $\frac{1}{2}$ of her money on a pack of markers, and $\frac{1}{8}$ of her money on a pack of pencils. What fraction of her money is left?

$1 - \frac{1}{3} - \frac{1}{2} - \frac{1}{8}$
$= \frac{2}{3} - \frac{1}{2} - \frac{1}{8}$
$= \frac{2}{3} - \frac{5}{8}$
$= \frac{16}{24} - \frac{15}{24}$
$= \frac{1}{24}$

Jing had $\frac{1}{24}$ of her money left.

3. Shelby bought a 2 ounce tube of blue paint. She used $\frac{2}{3}$ ounce to paint the water, $\frac{3}{5}$ ounce to paint the sky, and some to paint a flag. After that she has $\frac{2}{15}$ ounce left. How much paint did Shelby use to paint her flag?

$\frac{1}{3} + \frac{4}{15} = \frac{5}{15} + \frac{4}{15} = \frac{9}{15}$

Shelby used $\frac{9}{15}$ ounce to paint the flag.

T: Bring your Problem Set to the Debrief. Share, check, and/or explain your answers to your partner.

S: (Work together for 2 minutes.)

T: (Circulate and listen to students' explanations while they work, and then review answers.)

T: Let's take a look at two different strategies for solving Problem 4.

Solution 1

Jim ? | ? | $1\frac{5}{12}$ gallon

$1\frac{5}{12} - \frac{3}{4}$
$= \frac{1}{4} + \frac{5}{12}$
$= \frac{3}{12} + \frac{5}{12}$
$= \frac{8}{12}$ (Dwight's lemonade)

$\frac{3}{4} = \frac{9}{12}$

$\frac{9}{12} - \frac{8}{12} = \frac{1}{12}$

Jim sold $\frac{1}{12}$ gallon more than Dwight.

Solution 2

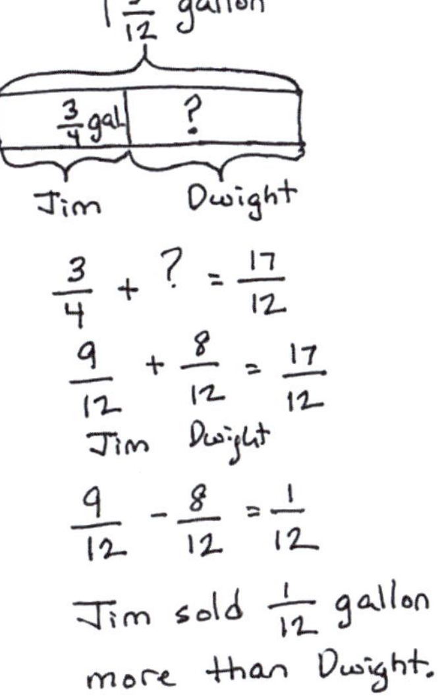

T: What do you notice about the 2 different drawings?

S: The first one shows the containers of lemonade, and the second one shows a tape diagram. → Both drawings are different, but they both have the part–part–whole relationship.

T: Let's look at them closely. How is Jim's container of $\frac{3}{4}$ gallon of lemonade represented in the tape diagram? Turn and share.

S: Instead of drawing a container of $\frac{3}{4}$ gallon, Jim's lemonade is now a part of a whole in the tape diagram.

T: How is Dwight's container of lemonade represented in the tape diagram? Turn and share.

S: (Share.) Since we don't know how much lemonade Dwight sold, we put a question mark in the container. But in the tape diagram, it's a missing part of a whole.

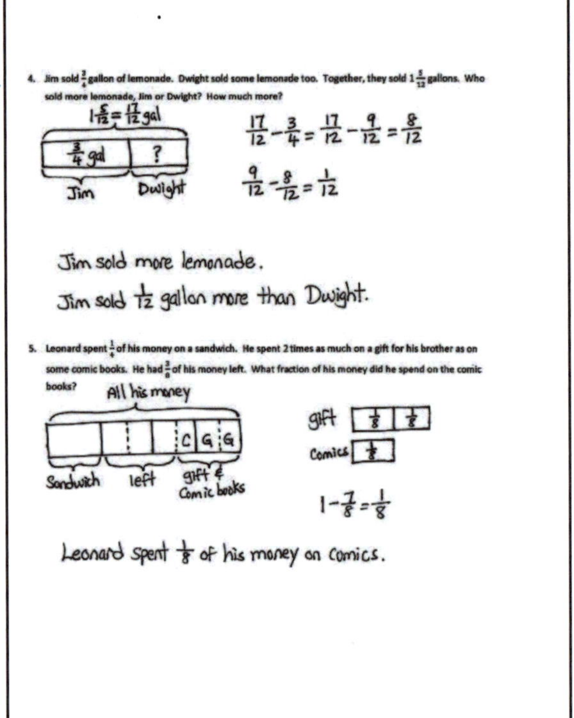
4. Jim sold $\frac{3}{4}$ gallon of lemonade. Dwight sold some lemonade too. Together, they sold $1\frac{5}{12}$ gallons. Who sold more lemonade, Jim or Dwight? How much more?

$1\frac{5}{12} = \frac{17}{12}$ gal

$\frac{3}{4}$ gal | ?

Jim Dwight

$\frac{17}{12} - \frac{3}{4} = \frac{17}{12} - \frac{9}{12} = \frac{8}{12}$

$\frac{9}{12} - \frac{8}{12} = \frac{1}{12}$

Jim sold more lemonade.

Jim sold $\frac{1}{12}$ gallon more than Dwight.

5. Leonard spent $\frac{1}{4}$ of his money on a sandwich. He spent 2 times as much on a gift for his brother as on some comic books. He had $\frac{3}{8}$ of his money left. What fraction of his money did he spend on the comic books?

All his money

C G G

Sandwich left gift & Comic books

gift $\frac{1}{8}$ $\frac{1}{8}$

Comics $\frac{1}{8}$

$1 - \frac{7}{8} = \frac{1}{8}$

Leonard spent $\frac{1}{8}$ of his money on comics.

T: Look at both drawings. How is the whole represented? Turn and share.

S: (Share.) The drawing on the left shows $1\frac{5}{12}$ gallons of lemonade in the containers. The drawing on the right shows the whole in a tape diagram.

MP.3

T: What if I change the numbers in this problem and make them larger? For example, Jim has $126\frac{3}{4}$ gallons, and the total is $348\frac{5}{12}$ gallons. Which drawing do you think is easier to draw and represent in the new problem? Turn and share.

S: (Share.) That's too many containers to draw. → It's easier to draw the new problem using the tape diagram. → It's faster to label the part–part–whole in the tape diagram than it is to draw all of the containers.

T: The tape diagram is much easier to use, especially with larger numbers.

T: What do you notice about their methods of solving this problem?

S: The second one started with the addition sentence $\frac{3}{4} + ? = 1\frac{5}{12}$, but the first one started with the subtraction sentence $1\frac{5}{12} - \frac{3}{4} = ?$.

NOTES ON MULTIPLE MEANS OF REPRESENTATION:

Ensure English language learners are sitting close to the displayed work during the Student Debrief, so that even when they lose the conversation's thread, they can analyze and learn from the student work.

English language learners may prefer the more concrete drawing of the gallons of lemonade. The explicit connection between the two representations is a strong bridge to understanding the tape diagram.

T: Turn and share with your partner, and follow each solution strategy step by step. Share what is the same and different about them.

S: (Share.)

T: If you have to solve a similar problem again, what kind of drawing and solution strategy would you use? Turn and share.

S: (Share.)

Exit Ticket (3 minutes)

After the Student Debrief, instruct students to complete the Exit Ticket. A review of their work will help with assessing students' understanding of the concepts that were presented in today's lesson and planning more effectively for future lessons. The questions may be read aloud to the students.

A

Number Correct: ______

Circle the Equivalent Fraction

1.	$2/4 =$	$1/2$	$1/3$	
2.	$2/6 =$	$1/2$	$1/3$	
3.	$2/8 =$	$1/2$	$1/4$	
4.	$5/10 =$	$1/2$	$1/4$	
5.	$5/15 =$	$1/2$	$1/3$	
6.	$5/20 =$	$1/2$	$1/4$	
7.	$4/8 =$	$1/2$	$1/4$	
8.	$4/12 =$	$1/2$	$1/3$	
9.	$4/16 =$	$1/2$	$1/4$	
10.	$3/6 =$	$1/2$	$1/3$	
11.	$3/9 =$	$1/2$	$1/3$	
12.	$3/12 =$	$1/2$	$1/4$	
13.	$4/6 =$	$2/3$	$1/3$	
14.	$6/12 =$	$2/3$	$1/2$	
15.	$6/18 =$	$2/3$	$1/3$	
16.	$6/30 =$	$1/5$	$1/3$	
17.	$6/9 =$	$2/3$	$1/3$	
18.	$7/14 =$	$1/2$	$1/3$	
19.	$7/21 =$	$1/2$	$1/3$	
20.	$7/42 =$	$1/6$	$1/7$	
21.	$8/12 =$	$2/3$	$3/4$	
22.	$9/18 =$	$1/2$	$1/3$	
23.	$9/27 =$	$2/3$	$1/3$	$1/4$
24.	$9/63 =$	$1/6$	$1/7$	$1/8$
25.	$8/12 =$	$2/3$	$3/4$	$4/5$
26.	$8/16 =$	$1/2$	$1/3$	$1/4$
27.	$8/24 =$	$1/2$	$1/3$	$1/4$
28.	$8/64 =$	$1/7$	$1/8$	$1/9$
29.	$12/18 =$	$3/4$	$5/6$	$2/3$
30.	$12/16 =$	$3/4$	$5/6$	$2/3$
31.	$9/12 =$	$3/4$	$5/6$	$2/3$
32.	$6/8 =$	$3/4$	$5/6$	$2/3$
33.	$10/12 =$	$3/4$	$5/6$	$2/3$
34.	$15/18 =$	$3/4$	$5/6$	$2/3$
35.	$8/10 =$	$3/4$	$4/5$	$2/3$
36.	$16/20 =$	$3/4$	$4/5$	$2/3$
37.	$12/15 =$	$3/4$	$4/5$	$2/3$
38.	$18/27 =$	$3/4$	$4/5$	$2/3$
39.	$27/36 =$	$3/4$	$4/5$	$2/3$
40.	$32/40 =$	$3/4$	$4/5$	$2/3$
41.	$45/54 =$	$3/4$	$4/5$	$5/6$
42.	$24/36 =$	$3/4$	$4/5$	$2/3$
43.	$60/72 =$	$3/4$	$5/6$	$2/3$
44.	$48/60 =$	$3/4$	$4/5$	$5/6$

B

Number Correct: _______

Improvement: _______

Circle the Equivalent Fraction

1.	$^5/_{10}$ =	$^1/_2$	$^1/_3$
2.	$^5/_{15}$ =	$^1/_2$	$^1/_3$
3.	$^5/_{20}$ =	$^1/_2$	$^1/_4$
4.	$^2/_4$ =	$^1/_2$	$^1/_3$
5.	$^2/_6$ =	$^1/_2$	$^1/_3$
6.	$^2/_8$ =	$^1/_2$	$^1/_4$
7.	$^3/_6$ =	$^1/_2$	$^1/_3$
8.	$^3/_9$ =	$^1/_2$	$^1/_3$
9.	$^3/_{12}$ =	$^1/_4$	$^1/_3$
10.	$^4/_8$ =	$^1/_2$	$^1/_3$
11.	$^4/_{12}$ =	$^1/_2$	$^1/_3$
12.	$^4/_{16}$ =	$^1/_4$	$^1/_3$
13.	$^4/_6$ =	$^2/_3$	$^1/_2$
14.	$^7/_{14}$ =	$^2/_3$	$^1/_2$
15.	$^7/_{21}$ =	$^1/_5$	$^1/_3$
16.	$^7/_{35}$ =	$^1/_5$	$^1/_3$
17.	$^6/_9$ =	$^2/_3$	$^1/_3$
18.	$^6/_{12}$ =	$^1/_2$	$^1/_3$
19.	$^6/_{18}$ =	$^1/_6$	$^1/_3$
20.	$^6/_{36}$ =	$^1/_6$	$^1/_3$
21.	$^8/_{12}$ =	$^2/_3$	$^3/_4$
22.	$^8/_{16}$ =	$^1/_2$	$^1/_3$

23.	$^8/_{24}$ =	$^2/_3$	$^1/_3$	$^1/_4$
24.	$^8/_{56}$ =	$^1/_6$	$^1/_7$	$^1/_8$
25.	$^8/_{12}$ =	$^2/_3$	$^3/_4$	$^4/_5$
26.	$^9/_{18}$ =	$^1/_2$	$^1/_3$	$^1/_4$
27.	$^9/_{27}$ =	$^1/_2$	$^1/_3$	$^1/_4$
28.	$^9/_{72}$ =	$^1/_7$	$^1/_8$	$^1/_9$
29.	$^{12}/_{18}$ =	$^3/_4$	$^5/_6$	$^2/_3$
30.	$^6/_8$ =	$^3/_4$	$^5/_6$	$^2/_3$
31.	$^9/_{12}$ =	$^3/_4$	$^5/_6$	$^2/_3$
32.	$^{12}/_{16}$ =	$^3/_4$	$^5/_6$	$^2/_3$
33.	$^8/_{10}$ =	$^3/_4$	$^4/_5$	$^2/_3$
34.	$^{16}/_{20}$ =	$^3/_4$	$^4/_5$	$^2/_3$
35.	$^{12}/_{15}$ =	$^3/_4$	$^4/_5$	$^2/_3$
36.	$^{10}/_{12}$ =	$^3/_4$	$^4/_5$	$^5/_6$
37.	$^{15}/_{18}$ =	$^3/_4$	$^5/_6$	$^2/_3$
38.	$^{16}/_{24}$ =	$^3/_4$	$^4/_5$	$^2/_3$
39.	$^{24}/_{32}$ =	$^3/_4$	$^4/_5$	$^2/_3$
40.	$^{36}/_{45}$ =	$^3/_4$	$^4/_5$	$^2/_3$
41.	$^{40}/_{48}$ =	$^3/_4$	$^4/_5$	$^5/_6$
42.	$^{24}/_{36}$ =	$^3/_4$	$^4/_5$	$^2/_3$
43.	$^{48}/_{60}$ =	$^3/_4$	$^5/_6$	$^4/_5$
44.	$^{60}/_{72}$ =	$^3/_4$	$^5/_6$	$^2/_3$

Name ________________________________ Date ________________

Solve the word problems using the RDW strategy. Show all of your work.

1. George weeded $\frac{1}{5}$ of the garden, and Summer weeded some, too. When they were finished, $\frac{2}{3}$ of the garden still needed to be weeded. What fraction of the garden did Summer weed?

2. Jing spent $\frac{1}{3}$ of her money on a pack of pens, $\frac{1}{2}$ of her money on a pack of markers, and $\frac{1}{8}$ of her money on a pack of pencils. What fraction of her money is left?

3. Shelby bought a 2-ounce tube of blue paint. She used $\frac{2}{3}$ ounce to paint the water, $\frac{3}{5}$ ounce to paint the sky, and some to paint a flag. After that, she has $\frac{2}{15}$ ounce left. How much paint did Shelby use to paint her flag?

4. Jim sold $\frac{3}{4}$ gallon of lemonade. Dwight sold some lemonade, too. Together, they sold $1\frac{5}{12}$ gallons. Who sold more lemonade, Jim or Dwight? How much more?

5. Leonard spent $\frac{1}{4}$ of his money on a sandwich. He spent 2 times as much on a gift for his brother as on some comic books. He had $\frac{3}{8}$ of his money left. What fraction of his money did he spend on the comic books?

Name ____ Date ____________________

Solve the word problem using the RDW strategy. Show all of your work.

Mr. Pham mowed $\frac{2}{7}$ of his lawn. His son mowed $\frac{1}{4}$ of it. Who mowed the most? How much of the lawn still needs to be mowed?

Name ______________________________ Date ______________

Solve the word problems using the RDW strategy. Show all of your work.

1. Christine baked a pumpkin pie. She ate $\frac{1}{6}$ of the pie. Her brother ate $\frac{1}{3}$ of it and gave the leftovers to his friends. What fraction of the pie did he give to his friends?

2. Liang went to the bookstore. He spent $\frac{1}{3}$ of his money on a pen and $\frac{4}{7}$ of it on books. What fraction of his money did he have left?

3. Tiffany bought $\frac{2}{5}$ kg of cherries. Linda bought $\frac{1}{10}$ kg of cherries less than Tiffany. How many kilograms of cherries did they buy altogether?

4. Mr. Rivas bought a can of paint. He used $\frac{3}{8}$ of it to paint a bookshelf. He used $\frac{1}{4}$ of it to paint a wagon. He used some of it to paint a birdhouse and has $\frac{1}{8}$ of the paint left. How much paint did he use for the birdhouse?

5. Ribbon A is $\frac{1}{3}$ m long. It is $\frac{2}{5}$ m shorter than Ribbon B. What's the total length of the two ribbons?

Name ______________________________ Date ________________

1. Lila collected the honey from 3 of her beehives. From the first hive she collected $\frac{2}{3}$ gallon of honey. The last two hives yielded $\frac{1}{4}$ gallon each.

   a. How many gallons of honey did Lila collect in all? Draw a diagram to support your answer.

   b. After using some of the honey she collected for baking, Lila found that she only had $\frac{3}{4}$ gallon of honey left. How much honey did she use for baking? Support your answer using a diagram, numbers, and words.

c. With the remaining $\frac{3}{4}$ gallon of honey, Lila decided to bake some loaves of bread and several batches of cookies for her school bake sale. The bread needed $\frac{1}{6}$ gallon of honey and the cookies needed $\frac{1}{4}$ gallon. How much honey was left over? Support your answer using a diagram, numbers, and words.

d. Lila decided to make more baked goods for the bake sale. She used $\frac{1}{8}$ lb less flour to make bread than to make cookies. She used $\frac{1}{4}$ lb more flour to make cookies than to make brownies. If she used $\frac{1}{2}$ lb of flour to make the bread, how much flour did she use to make the brownies? Explain your answer using a diagram, numbers, and words.

EUREKA MATH™

Mid-Module Assessment Task Standards Addressed	Topics A–B
Use equivalent fractions as a strategy to add and subtract fractions.	
5.NF.1	Add and subtract fractions with unlike denominators (including mixed numbers) by replacing given fractions with equivalent fractions in such a way as to produce an equivalent sum or difference of fractions with like denominators. *For example, 2/3 + 5/4 = 8/12 + 15/12 = 23/12. (In general, a/b + c/d = (ad + bc)/bd.)*
5.NF.2	Solve word problems involving addition and subtraction of fractions referring to the same whole, including cases of unlike denominators, e.g., by using visual fraction models or equations to represent the problem. Use benchmark fractions and number sense of fractions to estimate mentally and assess the reasonableness of answers. *For example, recognize an incorrect result 2/5 + 1/2 = 3/7, by observing that 3/7 < 1/2.*

Evaluating Student Learning Outcomes

A Progression Toward Mastery is provided to describe steps that illuminate the gradually increasing understandings that students develop on their way to proficiency. In this chart, this progress is presented from left (Step 1) to right (Step 4). The learning goal for students is to achieve Step 4 mastery. These steps are meant to help teachers and students identify and celebrate what the students CAN do now and what they need to work on next.

A Progression Toward Mastery				
Assessment Task Item and Standards Assessed	**STEP 1 Little evidence of reasoning without a correct answer.** **(1 Point)**	**STEP 2 Evidence of some reasoning without a correct answer.** **(2 Points)**	**STEP 3 Evidence of some reasoning with a correct answer or evidence of solid reasoning with an incorrect answer.** **(3 Points)**	**STEP 4 Evidence of solid reasoning with a correct answer.** **(4 Points)**
1(a) **5.NF.1**	Student shows little evidence of clear reasoning and understanding, resulting in an incorrect answer.	Student shows evidence of beginning to understand adding fractions with unlike denominators, but the answer is incorrect.	Student has the correct answer but is unable to show evidence accurately using diagrams, numbers, and/or words. OR Student shows evidence of correctly modeling adding of fractions with unlike denominators but results in an incorrect answer.	Student correctly: ▪ Calculates $\frac{14}{12}$ gal, $1\frac{2}{12}$ gal, $1\frac{1}{6}$ gal, $\frac{7}{6}$ gal, or equivalent. ▪ Illustrates the answer clearly in words, numbers, and a diagram.
1(b) **5.NF.1** **5.NF.2**	Student shows little evidence of using a correct strategy and understanding, resulting in the wrong answer.	Student shows evidence of beginning to understand subtracting fractions with unlike denominators but is unable to obtain the correct answer.	Student has the correct answer, but the model is either omitted or does not show evidence accurately using diagrams, numbers, and/or words. OR Student shows evidence of correctly modeling subtracting fractions with unlike denominators but results in an incorrect answer.	Student correctly: ▪ Calculates $\frac{5}{12}$ or $\frac{10}{24}$ gal. ▪ Illustrates the answer clearly in words, numbers, and a diagram.

A Progression Toward Mastery				
1(c) **5.NF.1** **5.NF.2**	Student shows little evidence of using a correct strategy and understanding, resulting in the wrong answer.	Student shows evidence of beginning to understand portions of the solution, such as attempting to add $\frac{1}{6}$ and $\frac{1}{4}$ and then subtract the result from $\frac{3}{4}$, but is unable to obtain the correct answer.	Student has the correct answer but the model is either omitted, or student is unable to show evidence accurately using diagrams, numbers, and/or words. OR Student shows evidence of correctly modeling adding and subtracting fractions with unlike denominators but results in an incorrect answer.	Student correctly: ▪ Calculates $\frac{1}{3}$ gal or equivalent fraction, such as $\frac{4}{12}$ gal. ▪ Models $\frac{1}{6}+\frac{1}{4}$ and $\frac{3}{4}-\frac{5}{12}$, or alternatively models $\frac{3}{4}-\frac{1}{6}-\frac{3}{4}$ using words, numbers, and a diagram.
1(d) **5.NF.1** **5.NF.2**	Student shows little evidence of using correct strategies, resulting in the wrong answer.	Student shows evidence of beginning to understand at least some of the steps involved but is unable to obtain the correct answer.	Student has the correct answer, but student does not show sound reasoning. OR Student demonstrates all steps using appropriate models but results in an incorrect answer.	Student correctly: ▪ Calculates $\frac{3}{8}$ lb as the amount of flour used for brownies. ▪ Diagrams and uses words and numbers to clearly explain the solution.

Name Jacqueline Date

1. Lila collected the honey from 3 of her beehives. From the first hive she collected $\frac{2}{3}$ gallon of honey. The last two hives yielded $\frac{1}{4}$ gallon each.

a. How many gallons of honey did Lila collect in all? Draw a diagram to support your answer.

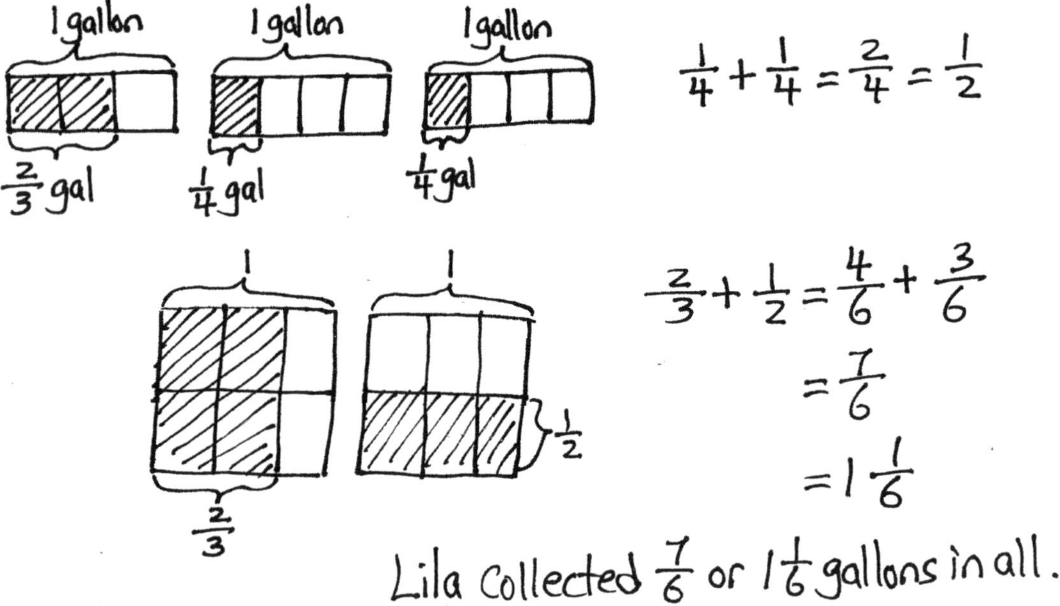

b. After using some of the honey she collected for baking, Lila found that she only had $\frac{3}{4}$ gallon of honey left. How much honey did she use for baking? Support your answer using a diagram, numbers, and words.

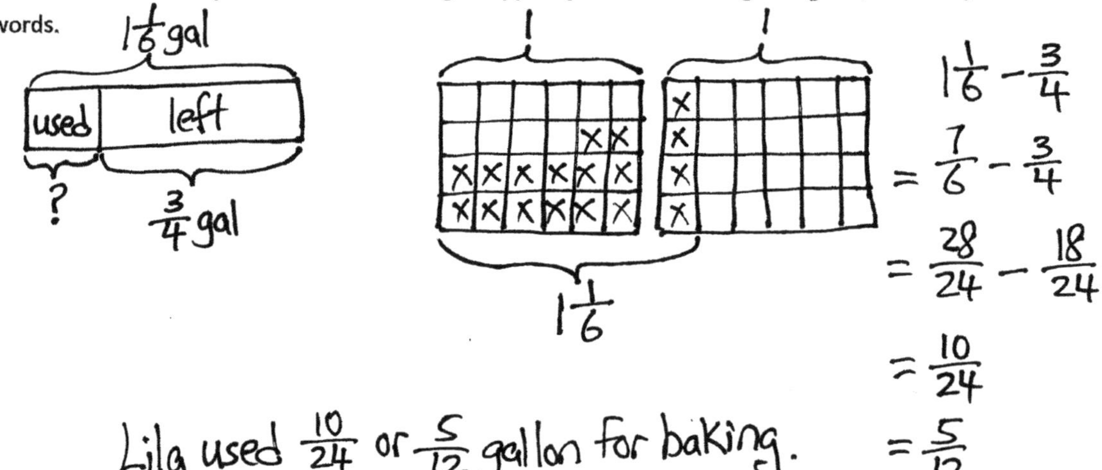

c. With the remaining $\frac{3}{4}$ gallon of honey, Lila decided to bake some loaves of bread and several batches of cookies for her school bake sale. The bread needed $\frac{1}{6}$ gallon of honey and the cookies needed $\frac{1}{4}$ gallon. How much honey was left over? Support your answer using a diagram, numbers, and words.

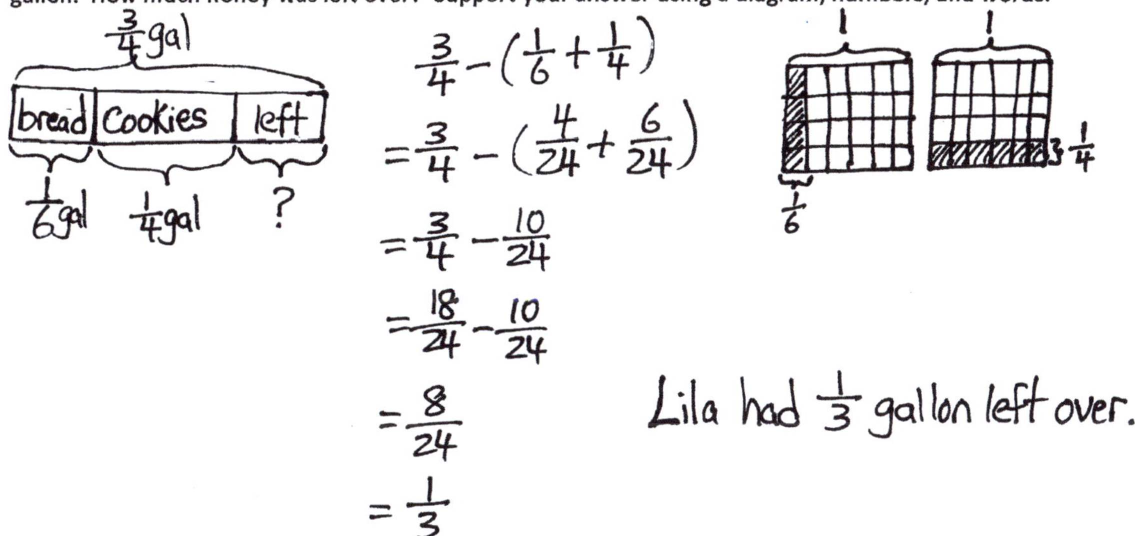

d. Lila decided to make more baked goods for the bake sale. She used $\frac{1}{8}$ lb less flour to make bread than to make cookies. She used $\frac{1}{4}$ lb more flour to make cookies than to make brownies. If she used $\frac{1}{2}$ lb of flour to make the bread, how much flour did she use to make the brownies? Explain your answer using a diagram, numbers, and words.

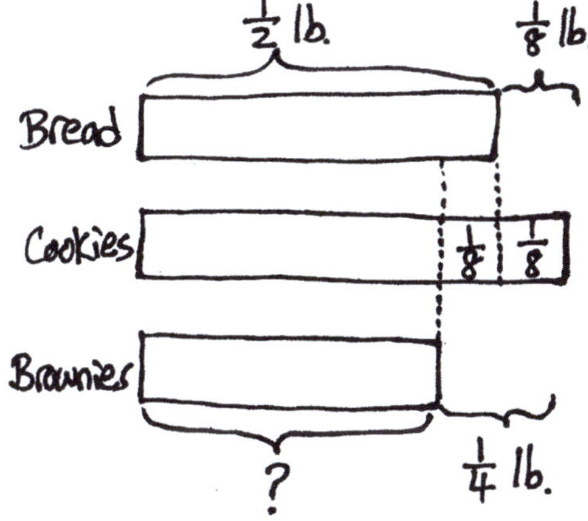

$\frac{1}{2}-\frac{1}{8}=\frac{4}{8}-\frac{1}{8}$

$=\frac{3}{8}$

Lila used $\frac{3}{8}$ pound of flour to make the brownies.

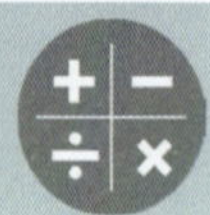

Topic C

Making Like Units Numerically

5.NF.1, 5.NF.2

Focus Standards:	5.NF.1	Add and subtract fractions with unlike denominators (including mixed numbers) by replacing given fractions with equivalent fractions in such a way as to produce an equivalent sum or difference of fractions with like denominators. *For example, 2/3 + 5/4 = 8/12 + 15/12 = 23/12. (In general, a/b + c/d = (ad + bc)/bd.)*
	5.NF.2	Solve word problems involving addition and subtraction of fractions referring to the same whole, including cases of unlike denominators, e.g., by using visual fraction models or equations to represent the problem. Use benchmark fractions and number sense of fractions to estimate mentally and assess the reasonableness of answers. *For example, recognize an incorrect result 2/5 + 1/2 = 3/7, by observing that 3/7 < 1/2.*
Instructional Days:	5	
Coherence -Links from:	G4–M5	Fraction Equivalence, Ordering, and Operations
-Links to:	G5–M1	Place Value and Decimal Fractions
	G5–M4	Multiplication and Division of Fractions and Decimal Fractions

In Topic C, students use the number line when adding and subtracting fractions greater than or equal to 1. The number line helps students see that fractions are analogous to whole numbers. The number line makes it clear that numbers on the left are smaller than numbers on the right, which leads to an understanding of integers in Grade 6. Using this tool, students recognize and manipulate fractions in relation to larger whole numbers and to each other. For example, "Between which two whole numbers does the sum of $1\frac{2}{3}$ and $5\frac{3}{4}$ lie?"

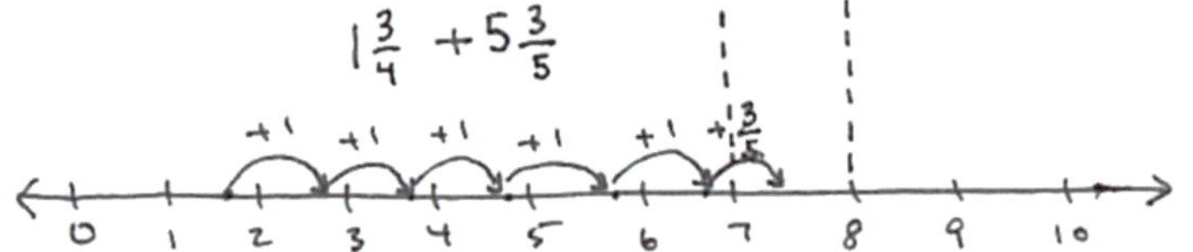

___ $< 1\frac{3}{4} + 5\frac{3}{5} <$ ___

This leads to an understanding of and skill with solving more complex problems often embedded within multi-step word problems:

Cristina and Matt's goal is to collect a total of $3\frac{1}{2}$ gallons of sap from the maple trees. Cristina collected $1\frac{3}{4}$ gallons. Matt collected $5\frac{3}{5}$ gallons. By how much did they beat their goal?

goal: $3\frac{1}{2}$ gal ?

collected: $1\frac{3}{4}$ gal $5\frac{3}{5}$ gal

$$1\frac{3}{4} + 5\frac{3}{5} - 3\frac{1}{2} = 3 + \left(\frac{3\times5}{4\times5}\right) + \left(\frac{3\times4}{5\times4}\right) - \left(\frac{1\times10}{2\times10}\right)$$

$$= 3 + \frac{15}{20} + \frac{12}{20} - \frac{10}{20}$$

$$= 3\,\frac{17}{20}$$

Cristina and Matt beat their goal by $3\frac{17}{20}$ gallons.

Word problems are a part of every lesson. Students are encouraged to utilize tape diagrams, which facilitate analysis of the same part–whole relationships they have worked with since Grade 1.

A Teaching Sequence Toward Mastery of Making Like Units Numerically

Objective 1: **Add fractions to and subtract fractions from whole numbers using equivalence and the number line as strategies.**
(Lesson 8)

Objective 2: **Add fractions making like units numerically.**
(Lesson 9)

Objective 3: **Add fractions with sums greater than 2.**
(Lesson 10)

Objective 4: **Subtract fractions making like units numerically.**
(Lesson 11)

Objective 5: **Subtract fractions greater than or equal to 1.**
(Lesson 12)

Lesson 8

Objective: Add fractions to and subtract fractions from whole numbers using equivalence and the number line as strategies.

Suggested Lesson Structure

Fluency Practice	(6 minutes)
Application Problem	(7 minutes)
Concept Development	(35 minutes)
Student Debrief	(12 minutes)
Total Time	**(60 minutes)**

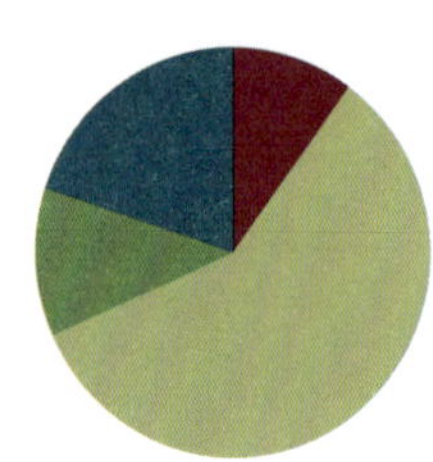

Fluency Practice (6 minutes)

- Adding Whole Numbers and Fractions **4.NF.3a** (3 minutes)
- Subtracting Fractions from Whole Numbers **4.NF.3a** (3 minutes)

Adding Whole Numbers and Fractions (3 minutes)

Note: This fluency activity reviews decomposing a mixed number into two addends—a whole number plus a fraction.

T: I'll say the answer. You say the addition problem as a whole number and a fraction. 3 and 1 half.
S: 3 + 1 half.
T: 5 and 1 half.
S: 5 + 1 half.
T: 2 and 3 fourths.
S: 2 + 3 fourths.
T: 1 and 5 sixths.
S: 1 + 5 sixths.
T: Let's switch roles. I'll say the addition problem. You say the answer. 2 + 1 fifth.
S: 2 and 1 fifth.
T: 2 + 4 fifths.
S: 2 and 4 fifths.
T: 5 + 7 eighths.
S: 5 and 7 eighths.
T: 3 + 7 twelfths.
S: 3 and 7 twelfths.

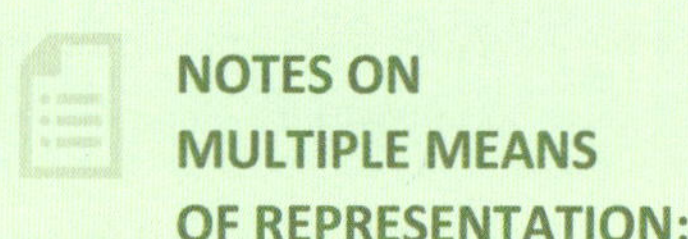

NOTES ON MULTIPLE MEANS OF REPRESENTATION:

If necessary, show numbers with tape diagrams to create a visual and slow the pace of the activity.

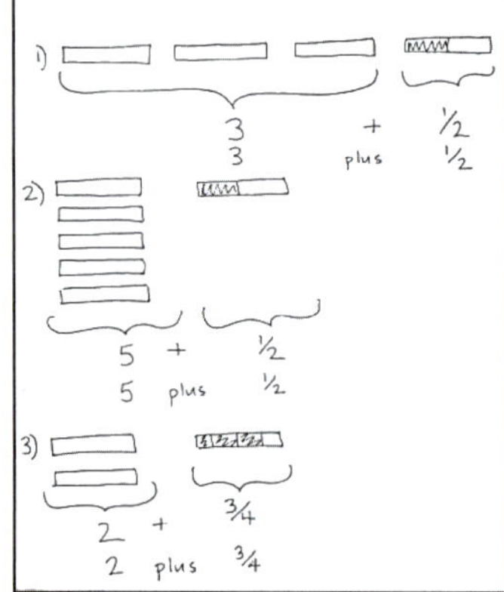

Subtracting Fractions from Whole Numbers (3 minutes)

Note: This fluency activity reviews subtraction of fractions. If students struggle with this activity, the problems can be written as shown in unit form.

T: I'll say a subtraction sentence. You repeat the sentence and give the answer. 1 – 1 half.

S: 1 – 1 half = 1 half.

T: 2 – 1 half.

S: 2 – 1 half = 1 and 1 half.

T: 2 and 1 half – 1 half.

S: 2 and 1 half – 1 half = 2.

T: 6 – 1 fourth.

S: 6 – 1 fourth = 5 and 3 fourths.

T: 6 and 3 fourths – 3 fourths.

S: 6 and 3 fourths – 3 fourths = 6.

Continue with the following possible sequence:

$3-\frac{5}{6}, 3\frac{5}{6}-\frac{5}{6}$ $\quad 4-\frac{7}{8}, 4\frac{7}{8}-\frac{7}{8}$ $\quad 5-\frac{7}{12}, 5\frac{7}{12}-\frac{7}{12}$.

NOTES ON MULTIPLE MEANS OF ACTION AND EXPRESSION:

As with the addition activity, have students represent the subtraction sentence with a tape diagram and cross off the subtracted amount.

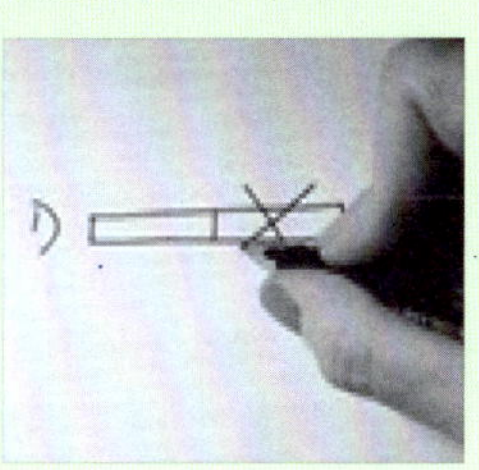

Application Problem (7 minutes)

Jane found money in her pocket. She went to a convenience store and spent $\frac{1}{4}$ of her money on chocolate milk, $\frac{3}{5}$ of her money on a magazine, and the rest of her money on candy. What fraction of her money did she spend on candy?

T: Let's read the problem together.

S: (Read chorally.)

T: Quickly share with your partner how to solve this problem. (Circulate and listen.)

T: What do we need to do to solve this problem?

S: I have to find like units for the cost of the milk and the magazine. Then, I can add them together. That way, I can see how much more I would need to make 1 whole.

T: You have 2 minutes to solve the problem.

T: What like units did you find for the milk and magazine?

S: Twentieths.

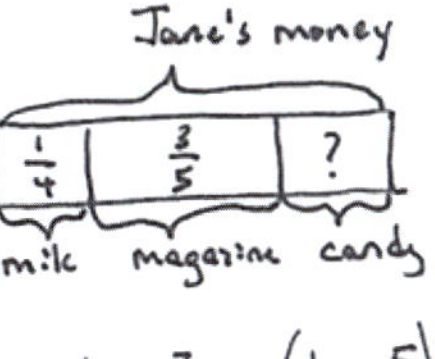

$\frac{1}{4}+\frac{3}{5}=\left(\frac{1}{4}\times\frac{5}{5}\right)+\left(\frac{3}{20}\times\frac{4}{4}\right)$

$=\frac{5}{20}+\frac{12}{20}$

$=\frac{17}{20}$

$\frac{17}{20}+?=\frac{20}{20}$

Jane spent $\frac{3}{20}$ of her money on candy.

T: Say your addition sentence with these like units.

S: 5 twentieths plus 12 twentieths equals 17 twentieths.

T: How many more twentieths do you need to make a whole?

S: 3 twentieths.

T: Tell your partner the answer in the form of a sentence.

S: Jane spent 3 twentieths of her money on candy.

Note: This Application Problem reviews addition of addends with unlike units.

> **NOTES ON MULTIPLE MEANS OF ENGAGEMENT:**
>
> Add the following question as an extension for students working above grade level:
>
> *How much does the magazine cost if she started with $10?*
>
> The question goes beyond the scope of the lesson but may be an engaging challenge for students working above grade level.

Concept Development (35 minutes)

Materials: (S) Personal white board, empty number line (Template) or lined paper

Problem 1: $\mathbf{1 + 1\frac{3}{4}}$

T: (Project or draw the image below.) If one fully shaded bar represents one whole, what addition problem would match this drawing?

S: $1 + 1\frac{3}{4}$.

T: (Write $1 + 1\frac{3}{4}$ on the board. Draw a line or project the number line template.) We'll start at zero and travel 1 unit. (Model.)

T: Start at 1 and travel one more equal unit. (Model.) Where do we land?

S: 2.

T: How much more do I need to add?

S: 3 fourths.

T: Will that additional distance be less than or more than one whole unit?

S: Less than one whole unit.

T: Make 3 smaller equal units: 1 fourth, 2 fourths,
3 fourths. What is 2 plus 3 fourths? Turn and share.

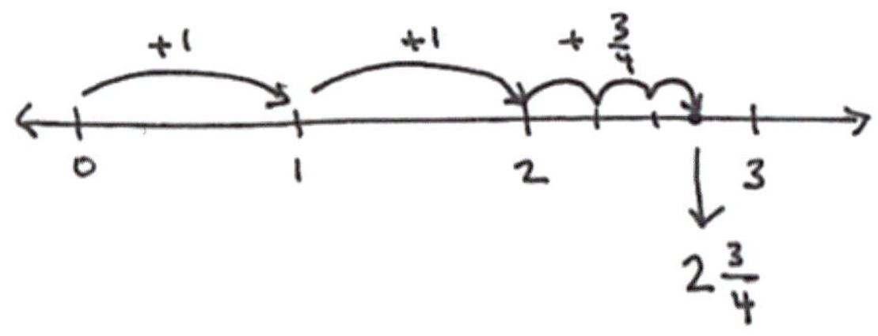

$$1 + 1\frac{3}{4}$$
$$= 1 + 1 + \frac{3}{4}$$
$$= 2\frac{3}{4}$$

S: 2 and 3 fourths. → 1 plus 1 and 3 fourths equals 2 and 3 fourths.

Problem 2: $2\frac{3}{10} + 3$

T: (Write $2\frac{3}{10} + 3$ on the board.) Talk to your partner. How should we solve this?

S: First, add 2. → 3 tenths comes next, so add that. → Adding all the whole numbers first might be easier. → Adding the numbers as they are written is best so you don't forget the fractions or whole numbers. → Adding the whole numbers first will make the number line easier to read, and it's similar to how we add all the ones, then the tens, then the hundreds. Add like numbers or units first.

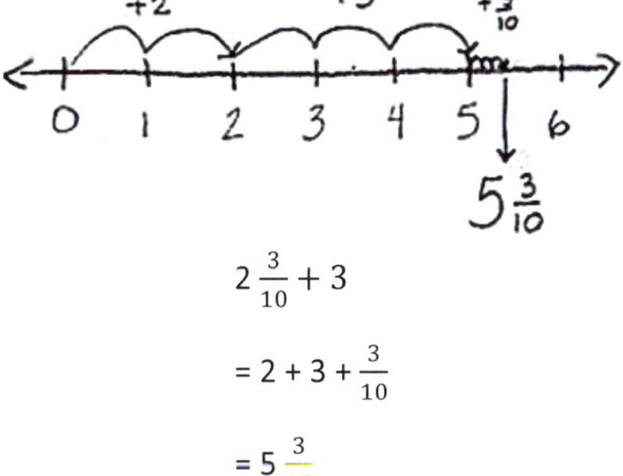

T: Let's travel 2 units and then 3 more units on our number line. (Show on the board.) Can someone explain how to travel 3 tenths?

S: 1 tenth is much smaller than a whole, so make 3 very small units. Label the final one $5\frac{3}{10}$.

T: Say your complete number sentence.

S: 2 and 3 tenths plus 3 equals 5 and 3 tenths.

T: What do you notice about the fractional units when adding them to a whole number?

S: The fraction amount doesn't change. All we have to do is add the whole numbers.

Problem 3: $1 - \frac{1}{4}$

T: (Write $1 - \frac{1}{4}$ on the board.) Read the problem.

S: 1 minus 1 fourth.

T: On the number line, let's start at 1 because that's the whole.

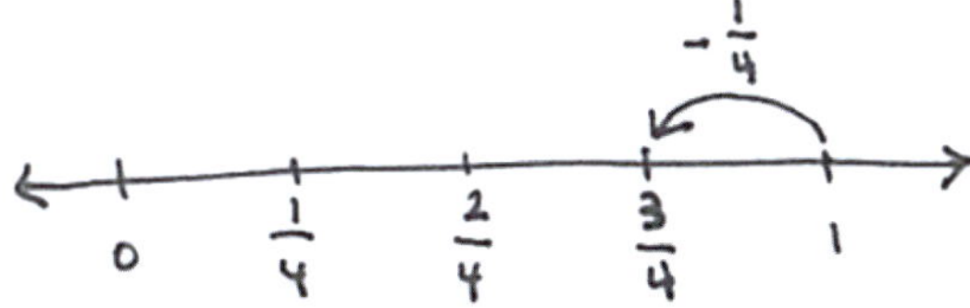

T: When I subtract $\frac{1}{4}$ from 1, my answer is between whi... 2 whole numbers?

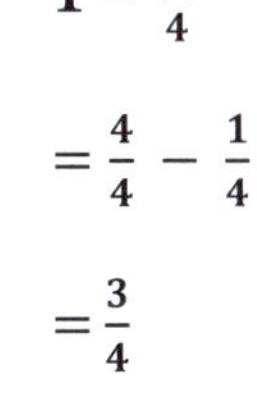

S: 0 and 1.

T: (Write 0 on the number line.) Because the answer is between 0 and 1, the whole number will be 0. Let's partition the number line into fourths. Starting at 1, let's travel back 1 fourth. (Mark the unit.) Say the complete number sentence.

S: 1 minus 1 fourth equals 3 fourths.

Problem 4: $2 - \frac{3}{5}$

T: (Write $2 - \frac{3}{5}$ on the board.) Discuss with your partner your strategy for solving this problem.

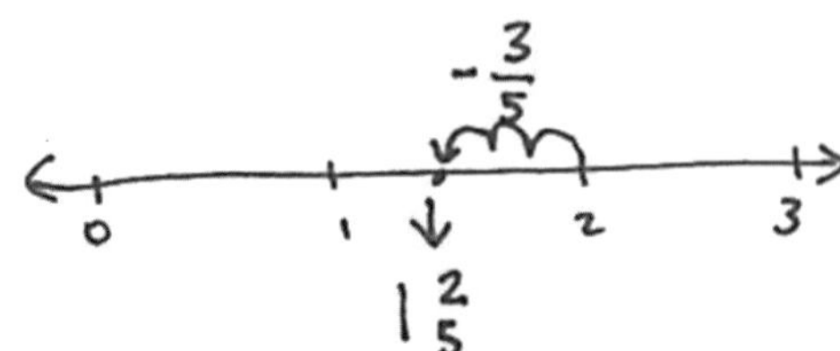

S: (Discuss.)

T: I will start at the whole number 2 on the number line. Am I subtracting a whole number?

S: No.

T: The answer will lie between what 2 whole numbers?

S: 1 and 2.

T: If the answer lies between 1 and 2, what is the whole number part of the answer?

S: 1.

T: With your partner, use your personal white board to subtract 3 fifths on the number line.

$$2 - \frac{3}{5}$$
$$= 1 + (1 - \frac{3}{5})$$
$$= 1\frac{2}{5}$$

Allow students 1 minute to solve the problem with their partners using the number line. Review the problem, counting back 3 fifths on the number line. Elicit the answer from students.

Problem 5: $3 - 1\frac{2}{3}$

T: (Write $3 - 1\frac{2}{3}$ on the board.) Say this subtraction sentence.

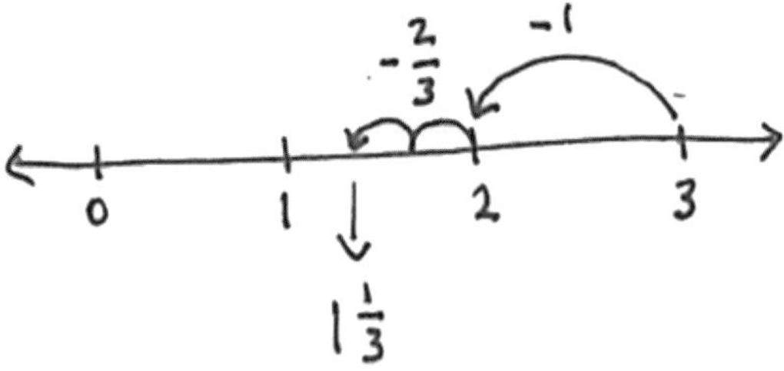

S: 3 minus 1 and 2 thirds.

T: First, we will subtract the whole number 1, and then subtract the fraction 2 thirds. Start with 3 on the number line, and subtract 1 whole. (Show the subtraction of the unit.)

T: When you subtract the fraction 2 thirds, what 2 whole numbers does the answer lie between?

S: Between 1 and 2.

T: You have 1 minute to complete this problem with your partner.

$$3 - 1\frac{2}{3}$$
$$= (3 - 1) - \frac{2}{3}$$
$$= 2 - \frac{2}{3}$$
$$= 1\frac{1}{3}$$

Problem Set (10 minutes)

Students should do their personal best to complete the Problem Set within the allotted 10 minutes. For some classes, it may be appropriate to modify the assignment by specifying which problems they work on first. Some problems do not specify a method for solving. Students should solve these problems using the RDW approach used for Application Problems.

Student Debrief (12 minutes)

Lesson Objective: Add fractions to and subtract fractions from whole numbers using equivalence and the number line as strategies.

The Student Debrief is intended to invite reflection and active processing of the total lesson experience.

Invite students to review their solutions for the Problem Set. They should check work by comparing answers with a partner before going over answers as a class. Look for misconceptions or misunderstandings that can be addressed in the Debrief. Guide students in a conversation to debrief the Problem Set and process the lesson.

T: Please take two minutes to check your answers with your partner. Do not change any of your answers. (Allow students to work.)

T: I will say the addition or subtraction problem. Share your answers out loud to check your work. Problem 1(a), 2 plus 1 and 1 fifth equals?

S: 3 and 1 fifth.

Continue with the sequence.

T: Take the next two minutes to discuss the Problem Set with your partner. Did you notice anything new? Are there any patterns? (Students discuss. Circulate and listen for conversations that can be shared with the whole class.)

T: Student A, will you tell us what you noticed about Problem 1(c)?

S: I added the whole numbers and got 7, but then I realized that the fractions added up to 5 fifths. That's one whole, so I had to add that to 7 and got 8 for my answer.

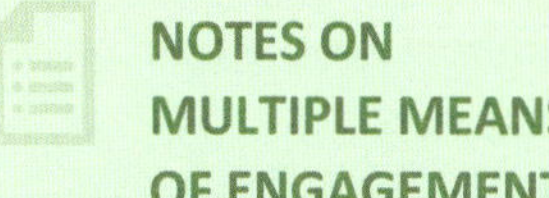

NOTES ON MULTIPLE MEANS OF ENGAGEMENT:

Learning how to articulate complex thinking is a skill that develops over time. It is often beneficial to have students project or show their work visually as they describe their solution strategies. Teachers can scaffold students' abilities to articulate their thoughts by questioning the projected strategies. Questions from classmates may also help students learn how to clearly articulate ideas. Ask students to retell particularly efficient strategies to a partner to help them internalize either language or content, depending on needs.

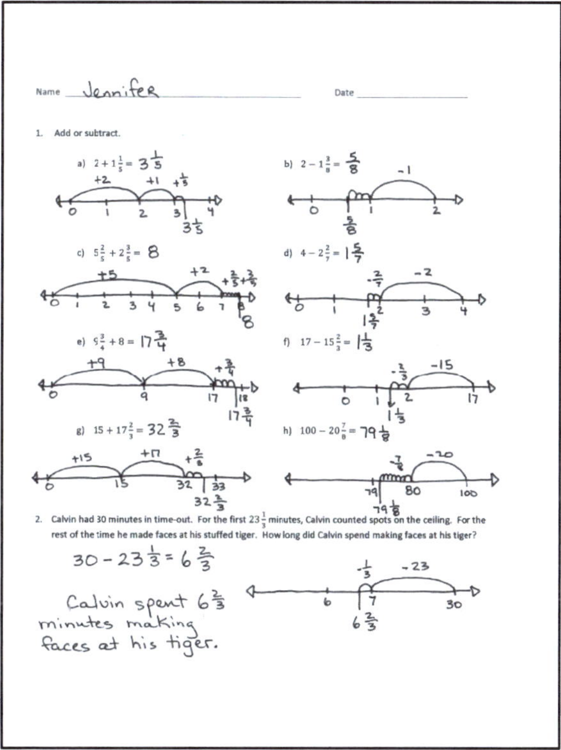

Name Jennifer Date

1. Add or subtract.

a) $2 + 1\frac{1}{5} = 3\frac{1}{5}$

b) $2 - 1\frac{3}{8} = \frac{5}{8}$

c) $5\frac{2}{5} + 2\frac{3}{5} = 8$

d) $4 - 2\frac{2}{7} = 1\frac{5}{7}$

e) $5\frac{3}{4} + 8 = 17\frac{3}{4}$

f) $17 - 15\frac{2}{3} = 1\frac{1}{3}$

g) $15 + 17\frac{2}{3} = 32\frac{2}{3}$

h) $100 - 20\frac{7}{8} = 79\frac{1}{8}$

2. Calvin had 30 minutes in time-out. For the first $23\frac{1}{3}$ minutes, Calvin counted spots on the ceiling. For the rest of the time he made faces at his stuffed tiger. How long did Calvin spend making faces at his tiger?

$30 - 23\frac{1}{3} = 6\frac{2}{3}$

Calvin spent $6\frac{2}{3}$ minutes making faces at his tiger.

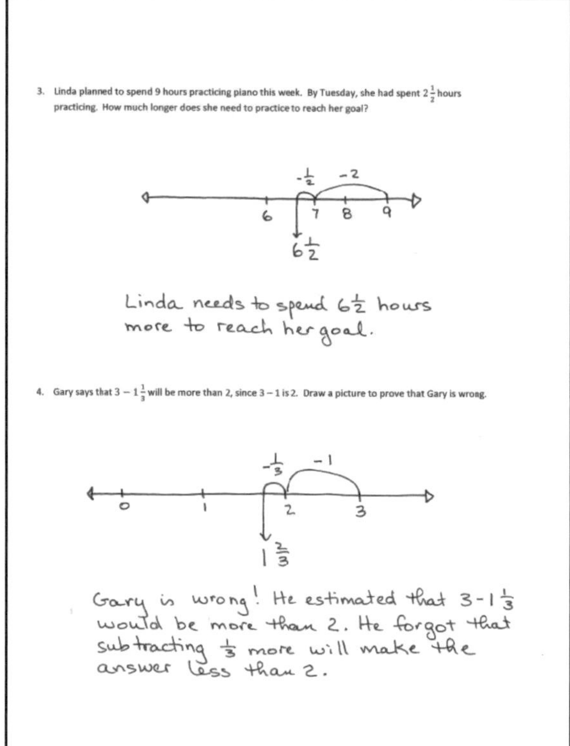

3. Linda planned to spend 9 hours practicing piano this week. By Tuesday, she had spent $2\frac{1}{2}$ hours practicing. How much longer does she need to practice to reach her goal?

Linda needs to spend $6\frac{1}{2}$ hours more to reach her goal.

4. Gary says that $3 - 1\frac{1}{3}$ will be more than 2, since $3 - 1$ is 2. Draw a picture to prove that Gary is wrong.

Gary is wrong! He estimated that $3 - 1\frac{1}{3}$ would be more than 2. He forgot that subtracting $\frac{1}{3}$ more will make the answer less than 2.

T: Student B, what were you saying about the addition problems compared to the subtraction problems?

S: Addition takes less time and thinking. Just add the whole numbers and write in the fraction. But with subtraction, you have to think harder. First, you subtract the whole numbers, but that won't be your whole number answer. You have to make it one number smaller. In Problem 1(f), for instance, 17 minus 15 equals 2, but the answer won't be 2; it will be between 1 and 2. So, I write down the whole number 1, and then figure out the fraction.

MP.3

T: Student C, how did you find the fraction that Student B mentioned?

S: For finding the fraction part of subtraction, I like to count up. For example, in Problem 1(d), I found the whole number and then said $\frac{3}{7}$, $\frac{4}{7}$, $\frac{5}{7}$, $\frac{6}{7}$, $\frac{7}{7}$. That's 5 groups of sevenths. So, the fraction is $\frac{5}{7}$.

T: Many of us are finding our own strategies for solving addition and subtraction of whole numbers and fractions. Share with your partner your own strategies. Listen carefully and see if you learn a new strategy to try.

S: (Discuss.)

T: (If time permits, ask two students to share what they heard.)

Exit Ticket (3 minutes)

After the Student Debrief, instruct students to complete the Exit Ticket. A review of their work will help with assessing students' understanding of the concepts that were presented in today's lesson and planning more effectively for future lessons. The questions may be read aloud to the students.

Name ______________________________ Date ________________

1. Add or subtract.

a. $2 + 1\frac{1}{5} =$

b. $2 - 1\frac{3}{8} =$

c. $5\frac{2}{5} + 2\frac{3}{5} =$

d. $4 - 2\frac{2}{7} =$

e. $9\frac{3}{4} + 8 =$

f. $17 - 15\frac{2}{3} =$

g. $15 + 17\frac{2}{3} =$

h. $100 - 20\frac{7}{8} =$

2. Calvin had 30 minutes in time-out. For the first $23\frac{1}{3}$ minutes, Calvin counted spots on the ceiling. For the rest of the time, he made faces at his stuffed tiger. How long did Calvin spend making faces at his tiger?

3. Linda planned to spend 9 hours practicing piano this week. By Tuesday, she had spent $2\frac{1}{2}$ hours practicing. How much longer does she need to practice to reach her goal?

4. Gary says that $3 - 1\frac{1}{3}$ will be more than 2, since $3 - 1$ is 2. Draw a picture to prove that Gary is wrong.

Name ______________________________ Date ______________

Add or subtract.

a. $5 + 1\frac{7}{8} =$

b. $3 - 1\frac{3}{4} =$

c. $7\frac{3}{8} + 4 =$

d. $4 - 2\frac{3}{7} =$

Name ______________________________ Date ______________

1. Add or subtract.

a. $3 + 1\frac{1}{4} =$

b. $2 - 1\frac{5}{8} =$

c. $5\frac{2}{5} + 2\frac{3}{5} =$

d. $4 - 2\frac{5}{7} =$

e. $8\frac{4}{5} + 7 =$

f. $18 - 15\frac{3}{4} =$

g. $16 + 18\frac{5}{6} =$

h. $100 - 50\frac{3}{8} =$

2. The total length of two ribbons is 13 meters. If one ribbon is $7\frac{5}{8}$ meters long, what is the length of the other ribbon?

3. It took Sandy two hours to jog 13 miles. She ran $7\frac{1}{2}$ miles in the first hour. How far did she run during the second hour?

4. Andre says that $5\frac{3}{4} + 2\frac{1}{4} = 7\frac{1}{2}$ because $7\frac{4}{8} = 7\frac{1}{2}$. Identify his mistake. Draw a picture to prove that he is wrong.

empty number line

Lesson 9

Objective: Add fractions making like units numerically.

Suggested Lesson Structure

Fluency Practice	(10 minutes)
Application Problem	(10 minutes)
Concept Development	(30 minutes)
Student Debrief	(10 minutes)
Total Time	**(60 minutes)**

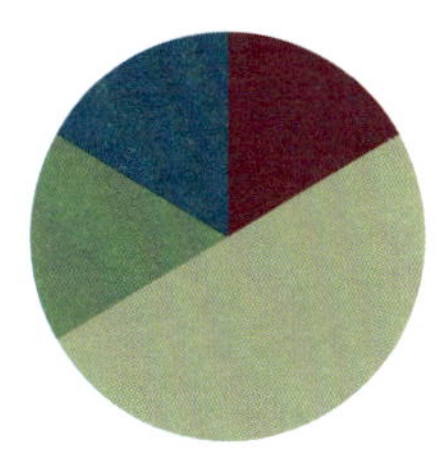

Fluency Practice (10 minutes)

- Adding and Subtracting Fractions with Like Units **4.NF.3a** (1 minute)
- Sprint: Add and Subtract Fractions with Like Units **4.NF.3a** (9 minutes)

Adding and Subtracting Fractions with Like Units (1 minute)

Note: This quick fluency activity reviews adding and subtracting like units mentally.

T: I'll say an addition or subtraction sentence. You say the answer. 2 fifths + 1 fifth.

S: 3 fifths.

T: 2 fifths – 1 fifth.

S: 1 fifth.

T: 2 fifths + 2 fifths.

S: 4 fifths.

T: 2 fifths – 2 fifths.

S: Zero.

T: 3 fifths + 2 fifths.

S: 1.

T: I'm going to write an addition sentence. You say whether it is true or false.

T: (Write $\frac{3}{7} + \frac{2}{7} = \frac{5}{7}$.)

S: True.

T: (Write $\frac{3}{7} + \frac{3}{7} = \frac{6}{14}$.)

S: False.

NOTES ON MULTIPLE MEANS OF REPRESENTATION:

Provide written equations alongside the oral presentation. Colored response cards (green represents true and red represents false) can help scaffold responses to the statement, "Tell me if it's true or false." This statement might also be simplified to "Is it right?" to which English language learners may respond "yes" or "no."

T: Say the answer that makes this addition sentence true.

S: 3 sevenths + 3 sevenths = 6 sevenths.

T: (Write $\frac{5}{9}+\frac{2}{9}=\frac{7}{18}$.)

T: True or false?

S: False.

T: Say the answer that makes this addition sentence true.

S: 5 ninths + 2 ninths = 7 ninths.

T: (Write $\frac{5}{9}+\frac{4}{9}=1$.)

T: True or false?

S: True.

T: Great work. You're ready for your Sprint!

NOTES ON MULTIPLE MEANS OF ENGAGEMENT:

The Application Problem may feel like review for some students. Consider extending it by asking, "If Hannah keeps to this training pattern, how many days will it take her to reach a distance of 2 miles?"

Also, consider having students generate other questions that could be asked about the story. For example:

- How far did Hannah run in 5 days?
- How much farther did Hannah run than her friend on Tuesday?
- How much farther did Hannah run on day 10 than day 1?

If students offer a question for which there is insufficient information, ask how the problem could be altered for their question to be answered.

Sprint: Add and Subtract Fractions with Like Units (9 minutes)

Materials: (S) Add and Subtract Fractions with Like Units Sprint

Note: This Sprint solidifies adding and subtracting fractions with like units and lays the groundwork for more advanced work with fractions.

Application Problem (10 minutes)

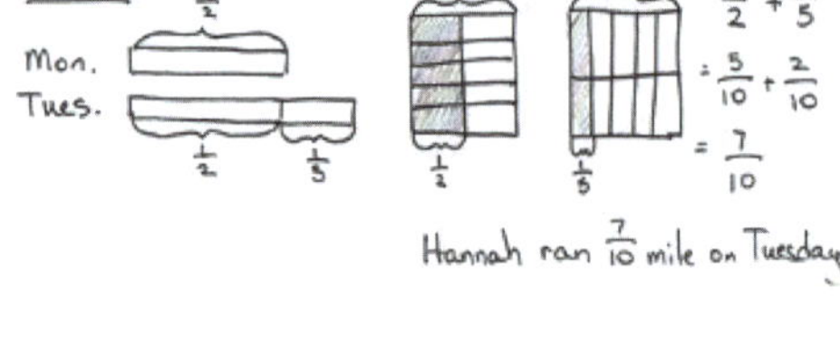

Hannah and her friend are training to run in a 2-mile race. On Monday, Hannah ran $\frac{1}{2}$ mile. On Tuesday, she ran $\frac{1}{5}$ mile farther than she ran on Monday.

a. How far did Hannah run on Tuesday?

b. If her friend ran $\frac{3}{4}$ mile on Tuesday, how many miles did the girls run in all on Tuesday?

Part B

Hannah $\frac{7}{10}$ Friend $\frac{3}{4}$

$\frac{7}{10}+\frac{3}{4}=$

$\frac{28}{40}+\frac{30}{40}=$

$\frac{58}{40}=\frac{40}{40}+\frac{18}{40}=$

$1\frac{18}{40}=1\frac{9}{20}$

In all, the girls ran $1\frac{9}{20}$ miles on Tuesday.

T: Use the RDW (read, draw, write) process to solve this problem with your partner.

S: (Read, draw, and write an equation, as well as a sentence to answer the question.)

T: (Debrief the problem.) Could you use the same units to answer Problems 1 (a) and (b)? Why or why not?

S: No. There's no easy way to change fourths to tenths.

Note: This Application Problem reviews addition of fractions with unlike denominators, using visual models as learned in earlier lessons. This pictorial strategy lays the foundation for a more abstract strategy for making like units introduced in this lesson's Concept Development.

Concept Development (30 minutes)

Materials: (S) Personal white board

Problem 1

> **NOTES ON MULTIPLE MEANS OF ENGAGEMENT:**
>
> The Turn and Talk strategy allows English language learners the opportunity to practice academic language in a relatively low-stakes setting. It also allows time for all students to formulate responses before sharing with the whole class.
>
> Consider pairing English language learners with each other at first, and then shift the language composition of groups over the course of the year.

T: How did you decide to use tenths in the first part of our Application Problem? Turn and talk.

S: We can draw a rectangle and split it using the other unit. → Since we had halves and fifths, we drew two parts and then split them into 5 parts each. That made 10 parts for the halves. That meant the fifths were each 2 smaller units, too.

T: Turn and talk. What happened to the number of units we selected when we split our rectangle?

S: Instead of one part, now we have five. → The number of selected parts is five times more. → The total number of parts is now 10.

T: What happened to the size of the units?

S: The units got smaller.

T: Let me record what I hear you saying. Does this equation say the same thing?

(Record the following equation.)

$$\left(\frac{1 \times 5}{2 \times 5}\right) = \left(\frac{5}{10}\right)$$

5 times as many selected units.
5 times as many units in the whole.

S: Yes!

T: Write an equation like mine to explain what happened to the fifths.

T: (Circulate and listen.) Jennifer, will you share for us?

S: The number of parts we had doubled. The units are half as big as before, but there are twice as many of them.

$$\left(\frac{1 \times 2}{5 \times 2}\right) = \frac{2}{10}$$

Number of parts doubled or 2 times as many parts.
Number of units in the whole doubled or twice as many parts in the whole.

T: Then, of course, we could add the two fractions together. (Write the equation as shown below.)

$$\left(\frac{1 \times 5}{2 \times 5}\right) + \left(\frac{1 \times 2}{5 \times 2}\right) = \frac{5}{10} + \frac{2}{10} = \frac{7}{10}$$

T: Are there other units we could have used to make these denominators the same? In other words, do 2 and 5 have other common multiples?

S: Yes. We could have used twentieths, thirtieths, or fiftieths.

T: If we had used twentieths, how many slices would we need to change $\frac{1}{2}$? To change $\frac{1}{5}$? Turn and talk. Draw a model on your personal white board, if necessary.

S: (Discuss.)

T: Let's hear your ideas.

S: 10 slices for half. → Ten times as many units in the whole, and 10 times as many units that were selected. → 4 slices for fifths. → 4 times as many selected units and 4 times as many units in the whole. The units are smaller in size.

T: Let's record that on our boards in equation form. (Write the equation as shown below.)

$$\left(\frac{1 \times 10}{2 \times 10}\right) + \left(\frac{1 \times 4}{5 \times 4}\right)$$
$$= \frac{10}{20} + \frac{4}{20}$$
$$= \frac{14}{20}$$

T: Is $\frac{14}{20}$ the same amount as $\frac{7}{10}$?

S: Yes, they are equivalent. → $\frac{7}{10}$ is simplified.

T: Express $\frac{1}{2} + \frac{1}{5}$ using another unit. Show your thinking with an equation.

S: (Draw and write appropriate representations.)

T: Who used the smallest unit? Who used the largest unit? Who had the least or most units in their whole? Turn and talk.

S: (Share.)

T: Please share your findings.

S: I used thirtieths, so I had to multiply both the numerator and denominator of $\frac{1}{2}$ by 15 and multiply both parts of $\frac{1}{5}$ by 6. That's $\frac{21}{30}$ in all. → I used fiftieths. I had smaller units in my whole, so I needed 25 to make $\frac{1}{2}$ and 10 to make $\frac{1}{5}$. That's $\frac{35}{50}$ in all.

T: (Write $\frac{1}{2} + \frac{1}{5} = \frac{7}{10} = \frac{14}{20} = \frac{21}{30} = \frac{35}{50}$.) Look at this equation. What do the types of units we used have in common?

S: All of the units are smaller than halves and fifths. → All are common multiples of 2 and 5. → All are multiples of ten.

T: Will the new unit always be a multiple of the original units? Think about this question as we solve the next problem.

Problem 2: $\frac{1}{2} + \frac{2}{3}$

T: (Write $\frac{1}{2} + \frac{2}{3}$.) How does this problem compare with our first problem?

S: It's still adding $\frac{1}{2}$ to something else. → The first problem was two unit fractions. → This problem only has one unit fraction. → We were adding an amount less than half to $\frac{1}{2}$ in the first problem, but $\frac{2}{3}$ is more than half.

$$\left(\frac{1 \times 3}{2 \times 3}\right) + \left(\frac{2 \times 2}{3 \times 2}\right)$$
$$= \frac{3}{6} + \frac{4}{6}$$
$$= \frac{7}{6} \text{ or } 1\frac{1}{6}$$

T: Great observations! What predictions can you make about the sum? For the units we use, what changes?

S: The sum should be a fraction greater than one. → We won't use most of the units from before.

EUREKA MATH™

T: How can you be sure?

S: We are adding half and more than half. → Only one of our units from before is a multiple of 3.

T: Imagine the rectangle that helps you find a like unit. Record an equation that explains what you saw in your mind's eye. (Circulate and observe.)

T: Show your equation, and explain it.

S: (Display equation.) I used sixths. My equation shows that for $\frac{1}{2}$, the number of pieces tripled, and the units in the whole tripled too. For $\frac{2}{3}$, the number of parts doubled, and so did the units in the whole.

T: Was our prediction about the answer correct?

S: Yes! The sum is greater than one!

T: Did anyone use another unit to find the sum? (Record sums on the board as students respond.)

T: Do these units follow the pattern we saw in our earlier work? Let's keep looking for evidence as we work.

NOTES ON MULTIPLE MEANS OF ACTION AND EXPRESSION:

While the focus of this lesson is the transition between pictorial and abstract representations of like units, allow students to continue to use the rectangular fraction model from previous lessons as a scaffold for writing and solving equations.

Problem 3: $\frac{5}{9}+\frac{5}{6}$

T: Compare this problem to the others. Turn and talk.

S: My partner and I see different things. I think this problem is like Problem 2 because the addends are more than half. But my partner says this one is like Problem 1 because the numerators are the same, even though they are not unit fractions. → This one is different because you can find a larger unit than the one you would use if you multiplied 6 and 9. → Eighteenths work as a like or common unit, and is a larger unit than fifty-fourths.

T: Find the sum. Use an equation to show your thinking.

$$\frac{5}{9}+\frac{5}{6}$$
$$=\left(\frac{5\times2}{9\times2}\right)+\left(\frac{5\times3}{6\times3}\right)$$
$$=\frac{10}{18}+\frac{15}{18}$$
$$=\frac{25}{18}=\frac{18}{18}+\frac{7}{18}=1\frac{7}{8}$$

Follow a similar procedure to Problems 1 and 2 to debrief the solution.

Problem 4: $\frac{2}{3}+\frac{1}{4}+\frac{1}{2}$

T: This problem has three addends. Does this affect our approach to solving?

S: No. We still have to find a like unit. It has to be a multiple of all three denominators.

T: Find the sum using an equation. (Debrief as above.)

T: To summarize, what patterns have you observed about the like units?

S: All the new units we found are common multiples of our original units. → We don't always have to multiply the original units to find a common multiple. → You can skip-count by the largest common unit, or like unit, to find smaller common or like units.

$$\frac{2}{3}+\frac{1}{4}+\frac{1}{2}$$
$$=\left(\frac{2\times4}{3\times4}\right)+\left(\frac{1\times3}{4\times3}\right)+\left(\frac{1\times6}{2\times6}\right)$$
$$=\frac{8}{12}+\frac{3}{12}+\frac{6}{12}$$
$$=\frac{17}{12}=\frac{12}{12}+\frac{5}{12}=1\frac{5}{12}$$

Problem Set (10 minutes)

Students should do their personal best to complete the Problem Set within the allotted 10 minutes. For some classes, it may be appropriate to modify the assignment by specifying which problems they work on first. Some problems do not specify a method for solving. Students should solve these problems using the RDW approach used for Application Problems.

Student Debrief (10 minutes)

Lesson Objective: Add fractions making like units numerically.

The Student Debrief is intended to invite reflection and active processing of the total lesson experience.

Invite students to review their solutions for the Problem Set. They should check work by comparing answers with a partner before going over answers as a class. Look for misconceptions or misunderstandings that can be addressed in the Debrief. Guide students in a conversation to debrief the Problem Set and process the lesson.

Any combination of the questions below may be used to lead the discussion.

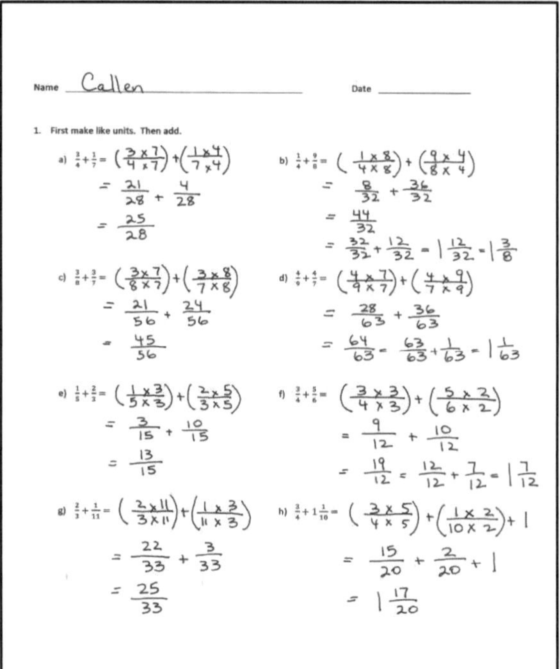
Name Callen Date

1. First make like units. Then add.

a) $\frac{3}{4}+\frac{1}{7}=\left(\frac{3\times7}{4\times7}\right)+\left(\frac{1\times4}{7\times4}\right)=\frac{21}{28}+\frac{4}{28}=\frac{25}{28}$

b) $\frac{1}{4}+\frac{9}{8}=\left(\frac{1\times8}{4\times8}\right)+\left(\frac{9\times4}{8\times4}\right)=\frac{8}{32}+\frac{36}{32}=\frac{44}{32}=\frac{32}{32}+\frac{12}{32}=1\frac{12}{32}=1\frac{3}{8}$

c) $\frac{3}{8}+\frac{3}{7}=\left(\frac{3\times7}{8\times7}\right)+\left(\frac{3\times8}{7\times8}\right)=\frac{21}{56}+\frac{24}{56}=\frac{45}{56}$

d) $\frac{4}{9}+\frac{4}{7}=\left(\frac{4\times7}{9\times7}\right)+\left(\frac{4\times9}{7\times9}\right)=\frac{28}{63}+\frac{36}{63}=\frac{64}{63}=\frac{63}{63}+\frac{1}{63}=1\frac{1}{63}$

e) $\frac{1}{5}+\frac{2}{3}=\left(\frac{1\times3}{5\times3}\right)+\left(\frac{2\times5}{3\times5}\right)=\frac{3}{15}+\frac{10}{15}=\frac{13}{15}$

f) $\frac{3}{4}+\frac{5}{6}=\left(\frac{3\times3}{4\times3}\right)+\left(\frac{5\times2}{6\times2}\right)=\frac{9}{12}+\frac{10}{12}=\frac{19}{12}=\frac{12}{12}+\frac{7}{12}=1\frac{7}{12}$

g) $\frac{2}{3}+\frac{1}{11}=\left(\frac{2\times11}{3\times11}\right)+\left(\frac{1\times3}{11\times3}\right)=\frac{22}{33}+\frac{3}{33}=\frac{25}{33}$

h) $\frac{3}{4}+1\frac{1}{10}=\left(\frac{3\times5}{4\times5}\right)+\left(\frac{1\times2}{10\times2}\right)+1=\frac{15}{20}+\frac{2}{20}+1=1\frac{17}{20}$

T: Check your answers with your partner. Please do not change any of them. (Allow time for students to confer.)

T: Did you notice any patterns in the sums on the Problem Set?

S: The answers in the first column were all less than a whole. → The answers in the second column were more than a whole.

T: I noticed that Problem 1(b) is different from all the other problems. Can you explain how it is different?

S: 1(b) is different because I only had to change the unit of one fraction to be like the other one. → One unit is a multiple of the other. → Eighths can be made out of fourths. None of the others were like that.

MP.7

T: Student A, please share your answer and your partner's answer to Problem 1(b).

S: I got 1 and $\frac{3}{8}$, but Student B got 1 and $\frac{12}{32}$.

T: Class, is it a problem that Student A's and Student B's answers to 1(b) are different?

S: No. It is the same amount. They just used different units. → You don't always have to multiply.

T: Did this situation come up more often in some problems than in others?

S: Yes. It happened more in Problems 1 (f) and (h).

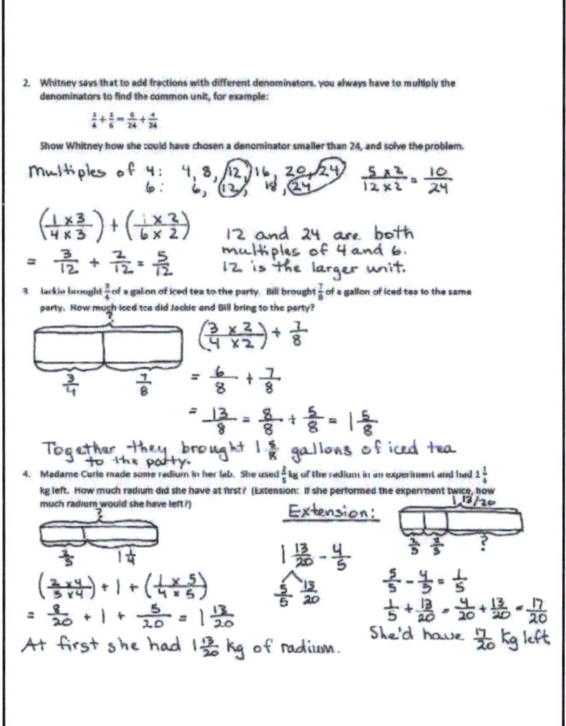

2. Whitney says that to add fractions with different denominators, you always have to multiply the denominators to find the common unit, for example:

$\frac{1}{4}+\frac{1}{6}=\frac{6}{24}+\frac{4}{24}$

Show Whitney how she could have chosen a denominator smaller than 24, and solve the problem.

Multiples of 4: 4, 8, 12, 16, 20, 24
6: 6, 12, 18, 24

$\frac{5\times2}{12\times2}=\frac{10}{24}$

$\left(\frac{1\times3}{4\times3}\right)+\left(\frac{1\times2}{6\times2}\right)$

$=\frac{3}{12}+\frac{2}{12}=\frac{5}{12}$

12 and 24 are both multiples of 4 and 6. 12 is the larger unit.

3. Jackie brought $\frac{3}{4}$ of a gallon of iced tea to the party. Bill brought $\frac{7}{8}$ of a gallon of iced tea to the same party. How much iced tea did Jackie and Bill bring to the party?

$\left(\frac{3\times2}{4\times2}\right)+\frac{7}{8}$

$=\frac{6}{8}+\frac{7}{8}$

$=\frac{13}{8}=\frac{8}{8}+\frac{5}{8}=1\frac{5}{8}$

Together they brought $1\frac{5}{8}$ gallons of iced tea to the party.

4. Madame Curie made some radium in her lab. She used $\frac{2}{5}$ kg of the radium in an experiment and had $1\frac{1}{4}$ kg left. How much radium did she have at first? (Extension: If she performed the experiment twice, how much radium would she have left?)

$\left(\frac{2\times4}{5\times4}\right)+1+\left(\frac{1\times5}{4\times5}\right)$

$=\frac{8}{20}+1+\frac{5}{20}=1\frac{13}{20}$

At first she had $1\frac{13}{20}$ kg of radium.

Extension:

$1\frac{13}{20}-\frac{4}{5}$

$\frac{5}{5}-\frac{4}{5}=\frac{1}{5}$

$\frac{1}{5}+\frac{13}{20}=\frac{4}{20}+\frac{13}{20}=\frac{17}{20}$

She'd have $\frac{17}{20}$ kg left.

MP.7

T: Why?

S: Multiplying the units together in these didn't give us the largest unit they had in common. → I could find a smaller common multiple than just multiplying them together. → I skip-counted by the smaller denominator until I got to a multiple of the other denominator. → If I multiplied them together, I could simplify the answer I got to use a larger unit.

T: How can these observations help you answer Problem 2?

S: Problem 2 was like Problems 1 (f) and (h). There was a larger unit in common. → You can slice the units by the same number to get a common unit.

T: Terrific insights! Put them to use as you complete your Exit Ticket.

Exit Ticket (3 minutes)

After the Student Debrief, instruct students to complete the Exit Ticket. A review of their work will help with assessing students' understanding of the concepts that were presented in today's lesson and planning more effectively for future lessons. The questions may be read aloud to the students.

A

Number Correct: _______

Add and Subtract Fractions with Like Units

1.	$\frac{1}{5}+\frac{1}{5}=$		23.	$\frac{1}{9}+\frac{1}{9}+\frac{1}{9}=$	
2.	$\frac{1}{10}+\frac{5}{10}=$		24.	$\frac{1}{9}+\frac{3}{9}+\frac{1}{9}=$	
3.	$\frac{1}{10}+\frac{7}{10}=$		25.	$\frac{4}{9}-\frac{1}{9}-\frac{3}{9}=$	
4.	$\frac{2}{5}+\frac{2}{5}=$		26.	$\frac{1}{4}+\frac{2}{4}+\frac{1}{4}=$	
5.	$\frac{5}{10}-\frac{4}{10}=$		27.	$\frac{1}{8}+\frac{3}{8}+\frac{2}{8}=$	
6.	$\frac{3}{5}-\frac{1}{5}=$		28.	$\frac{5}{12}+\frac{1}{12}+\frac{5}{12}=$	
7.	$\frac{3}{10}+\frac{3}{10}=$		29.	$\frac{2}{9}+\frac{3}{9}+\frac{2}{9}=$	
8.	$\frac{4}{5}-\frac{1}{5}=$		30.	$\frac{3}{10}-\frac{3}{10}+\frac{3}{10}=$	
9.	$\frac{1}{4}+\frac{1}{4}=$		31.	$\frac{3}{5}-\frac{1}{5}-\frac{1}{5}=$	
10.	$\frac{1}{4}+\frac{2}{4}=$		32.	$\frac{1}{6}+\frac{2}{6}=$	
11.	$\frac{3}{12}-\frac{2}{12}=$		33.	$\frac{3}{12}+\frac{4}{12}=$	
12.	$\frac{1}{4}+\frac{3}{4}=$		34.	$\frac{3}{12}+\frac{6}{12}=$	
13.	$\frac{1}{12}+\frac{1}{12}=$		35.	$\frac{4}{8}+\frac{2}{8}=$	
14.	$\frac{1}{3}+\frac{1}{3}=$		36.	$\frac{4}{12}+\frac{1}{12}=$	
15.	$\frac{3}{12}-\frac{2}{12}=$		37.	$\frac{1}{5}+\frac{3}{5}=$	
16.	$\frac{5}{12}+\frac{6}{12}=$		38.	$\frac{2}{5}+\frac{2}{5}=$	
17.	$\frac{7}{12}+\frac{4}{12}=$		39.	$\frac{1}{6}+\frac{2}{6}=$	
18.	$\frac{4}{6}-\frac{1}{6}=$		40.	$\frac{5}{12}-\frac{3}{12}=$	
19.	$\frac{1}{6}+\frac{2}{6}=$		41.	$\frac{7}{15}-\frac{2}{15}=$	
20.	$\frac{1}{6}+\frac{1}{6}+\frac{1}{6}=$		42.	$\frac{7}{15}-\frac{3}{15}=$	
21.	$\frac{1}{3}+\frac{1}{3}+\frac{1}{3}=$		43.	$\frac{11}{15}-\frac{2}{15}=$	
22.	$\frac{1}{12}+\frac{1}{12}+\frac{1}{12}=$		44.	$\frac{2}{15}+\frac{4}{15}=$	

B

Number Correct: _______

Improvement: _______

Add and Subtract Fractions with Like Units

1.	$\frac{1}{2}+\frac{1}{2}=$	
2.	$\frac{2}{8}+\frac{1}{8}=$	
3.	$\frac{2}{8}+\frac{3}{8}=$	
4.	$\frac{2}{12}-\frac{1}{12}=$	
5.	$\frac{5}{12}+\frac{2}{12}=$	
6.	$\frac{4}{8}+\frac{3}{8}=$	
7.	$\frac{4}{8}-\frac{3}{8}=$	
8.	$\frac{1}{8}+\frac{5}{8}=$	
9.	$\frac{3}{4}-\frac{1}{4}=$	
10.	$\frac{3}{6}-\frac{3}{6}=$	
11.	$\frac{3}{9}+\frac{3}{9}=$	
12.	$\frac{2}{3}+\frac{1}{3}=$	
13.	$\frac{6}{9}-\frac{4}{9}=$	
14.	$\frac{5}{9}-\frac{3}{9}=$	
15.	$\frac{2}{9}+\frac{2}{9}=$	
16.	$\frac{1}{12}+\frac{3}{12}=$	
17.	$\frac{5}{12}-\frac{4}{12}=$	
18.	$\frac{9}{12}-\frac{6}{12}=$	
19.	$\frac{6}{10}-\frac{4}{10}=$	
20.	$\frac{2}{8}+\frac{2}{8}+\frac{2}{8}=$	
21.	$\frac{1}{10}+\frac{1}{10}+\frac{1}{10}=$	
22.	$\frac{7}{10}-\frac{2}{10}-\frac{4}{10}=$	

23.	$\frac{1}{12}+\frac{6}{12}+\frac{2}{12}=$	
24.	$\frac{4}{12}+\frac{3}{12}+\frac{3}{12}=$	
25.	$\frac{8}{12}-\frac{4}{12}-\frac{4}{12}=$	
26.	$\frac{1}{10}+\frac{2}{10}+\frac{4}{10}=$	
27.	$\frac{1}{10}+\frac{1}{10}+\frac{6}{10}=$	
28.	$\frac{4}{6}+\frac{1}{6}+\frac{1}{6}=$	
29.	$\frac{2}{12}+\frac{3}{12}+\frac{4}{12}=$	
30.	$\frac{2}{10}+\frac{4}{10}+\frac{4}{10}=$	
31.	$\frac{3}{10}+\frac{1}{10}+\frac{2}{10}=$	
32.	$\frac{4}{6}-\frac{2}{6}=$	
33.	$\frac{3}{12}-\frac{2}{12}=$	
34.	$\frac{2}{3}+\frac{1}{3}=$	
35.	$\frac{2}{4}+\frac{1}{4}=$	
36.	$\frac{3}{12}+\frac{2}{12}=$	
37.	$\frac{1}{5}+\frac{2}{5}=$	
38.	$\frac{4}{5}-\frac{4}{5}=$	
39.	$\frac{5}{12}-\frac{1}{12}=$	
40.	$\frac{6}{8}+\frac{2}{8}=$	
41.	$\frac{2}{8}+\frac{2}{8}+\frac{2}{8}=$	
42.	$\frac{9}{10}-\frac{7}{10}-\frac{1}{10}=$	
43.	$\frac{2}{10}+\frac{5}{10}+\frac{2}{10}=$	
44.	$\frac{9}{12}-\frac{1}{12}-\frac{4}{12}=$	

Name ______________________________ Date ______________

1. First make like units, and then add.

a. $\frac{3}{4} + \frac{1}{7} =$

b. $\frac{1}{4} + \frac{9}{8} =$

c. $\frac{3}{8} + \frac{3}{7} =$

d. $\frac{4}{9} + \frac{4}{7} =$

e. $\frac{1}{5} + \frac{2}{3} =$

f. $\frac{3}{4} + \frac{5}{6} =$

g. $\frac{2}{3}+\frac{1}{11}=$

h. $\frac{3}{4}+1\frac{1}{10}=$

2. Whitney says that to add fractions with different denominators, you always have to multiply the denominators to find the common unit; for example:

$$\frac{1}{4}+\frac{1}{6}=\frac{6}{24}+\frac{4}{24}.$$

Show Whitney how she could have chosen a denominator smaller than 24, and solve the problem.

3. Jackie brought $\frac{3}{4}$ of a gallon of iced tea to the party. Bill brought $\frac{7}{8}$ of a gallon of iced tea to the same party. How much iced tea did Jackie and Bill bring to the party?

4. Madame Curie made some radium in her lab. She used $\frac{2}{5}$ kg of the radium in an experiment and had $1\frac{1}{4}$ kg left. How much radium did she have at first? (Extension: If she performed the experiment twice, how much radium would she have left?)

Name ______________________________ Date ______________

Make like units, and then add.

a. $\frac{1}{6} + \frac{3}{4} =$

b. $1\frac{1}{2} + \frac{2}{5} =$

Name ______________________________ Date ______________

1. Make like units, and then add.

a. $\frac{3}{5}+\frac{1}{3}=$

b. $\frac{3}{5}+\frac{1}{11}=$

c. $\frac{2}{9}+\frac{5}{6}=$

d. $\frac{2}{5}+\frac{1}{4}+\frac{1}{10}=$

e. $\frac{1}{3}+\frac{7}{5}=$

f. $\frac{5}{8}+\frac{7}{12}=$

g. $1\frac{1}{3} + \frac{3}{4} =$

h. $\frac{5}{6} + 1\frac{1}{4} =$

2. On Monday, Ka practiced guitar for $\frac{2}{3}$ of one hour. When she finished, she practiced piano for $\frac{3}{4}$ of one hour. How much time did Ka spend practicing instruments on Monday?

3. Ms. How bought a bag of rice for dinner. She used $\frac{3}{5}$ kg of the rice and still had $2\frac{1}{4}$ kg left. How heavy was the bag of rice that Ms. How bought?

4. Joe spends $\frac{2}{5}$ of his money on a jacket and $\frac{3}{8}$ of his money on a shirt. He spends the rest on a pair of pants. What fraction of his money does he use to buy the pants?

Lesson 10

Objective: Add fractions with sums greater than 2.

Suggested Lesson Structure

■	Fluency Practice	(10 minutes)
■	Application Problem	(8 minutes)
■	Concept Development	(32 minutes)
■	Student Debrief	(10 minutes)
	Total Time	**(60 minutes)**

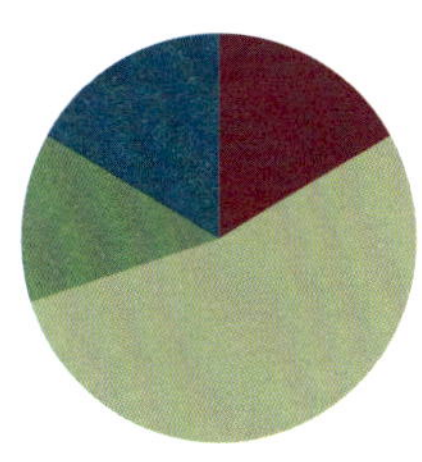

Fluency Practice (10 minutes)

- Sprint: Add and Subtract Whole Numbers and Ones with Fraction Units **4.NF.3c** (10 minutes)

Sprint: Add and Subtract Whole Numbers and Ones with Fraction Units (10 minutes)

Materials: (S) Add and Subtract Whole Numbers and Ones with Fraction Units Sprint

Note: This Sprint strengthens prerequisite skills for today's fractional work with sums greater than 2.

Application Problem (8 minutes)

To make punch for the class party, Mrs. Lui mixed $1\frac{1}{3}$ cups orange juice, $\frac{3}{4}$ cup apple juice, $\frac{2}{3}$ cup cranberry juice, and $\frac{3}{4}$ cup lemon-lime soda. Mixed together, how many cups of punch does the recipe make? (Extension: Each serving is 1 cup. How many batches of this recipe does Mrs. Lui need to serve her 20 students?)

T: Let's read the problem together.

S: (Read chorally.)

T: Can you draw something? Use the RDW process to solve the problem. (Circulate while students work.)

T: Alexis, will you tell the class about your solution?

S: I noticed that Mrs. Lui uses thirds and fourths when measuring. I added the like units together first. Then, I added the unlike units last to find the answer.

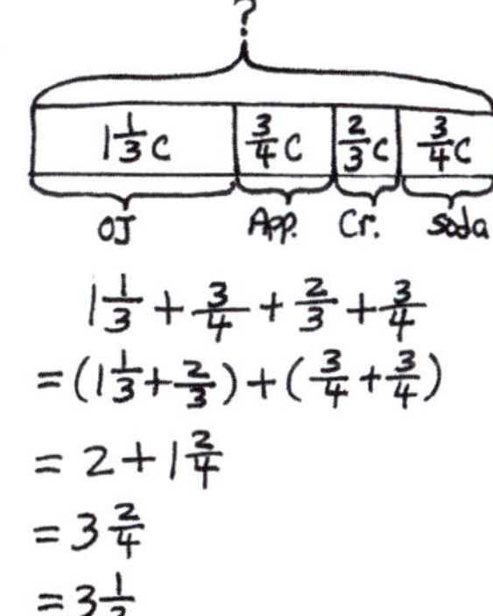

T: Say the addition sentence for the units of thirds.

S: $1\frac{1}{3} + \frac{2}{3} = 2.$

T: 2 what?

S: 2 cups.

T: Say your addition sentence for the units of fourths.

S: 3 fourths + 3 fourths = 1 and 1 half.

T: 1 and 1 half what?

S: 1 and 1 half cups.

T: How do I finish solving this problem?

S: Add 2 cups + 1 and 1 half cups.

T: Tell your partner your final answer as a sentence.

S: Mrs. Lui's recipe makes 3 and 1 half cups of punch.

NOTES ON MULTIPLE MEANS OF REPRESENTATION:

Every so often during fraction work, students discuss the need for like units when adding or subtracting. Remind them that this is also true with whole numbers. When they add or subtract whole numbers, they add or subtract ones with ones, tens with tens, hundreds with hundreds, etc. It is very important for students to clearly understand that fractions follow the same rules as whole numbers. Like units are always necessary to add or subtract.

If time allows, ask students to share strategies for solving the extension question.

Note: This Application Problem reviews Topic B skills, particularly adding unlike fractions from numbers between 0 and 2, in preparation for today's addition of fractions with sums greater than 2.

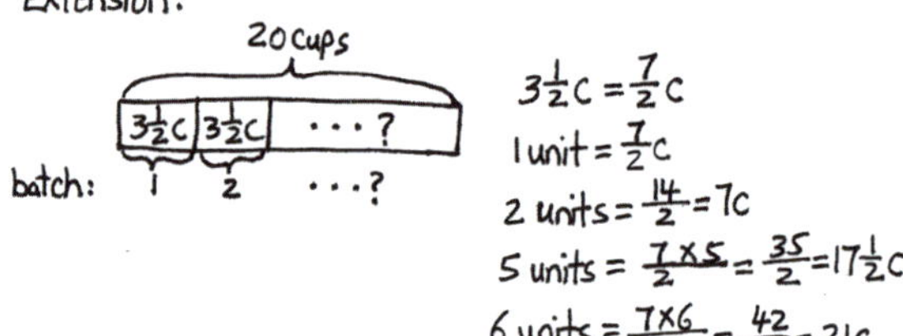

Concept Development (32 minutes)

Materials: (S) Personal white board

T: (Post expressions A, B, and C on the board.) Look at the three expressions on the board. Discuss with your partner how they are similar and how they are different.

A: $2\frac{1}{5} + 1\frac{1}{5}$ B: $2\frac{1}{5} + 1\frac{1}{2}$ C: $2\frac{4}{5} + 1\frac{1}{2}$

S: Each of the expressions adds whole numbers plus fractional units. → The fractions in expression A have like units, fifths. → The fractional units are different in B and C. → Both A and B will result in an answer between 3 and 4. The sum for expression C will be between 4 and 5.

NOTES ON MULTIPLE MEANS OF ENGAGEMENT:

Throughout this lesson, students are asked to work with fraction equations and understand how each expression within the equation progresses to the next. Have English language learners who share a first language sit together to comfortably discuss the analysis of the fraction equalities.

Problem 1

T: Read expression B.

S: 2 and $\frac{1}{5}$ + 1 and $\frac{1}{2}$.

T: Discuss with your partner if the following equation is true. (Write expressions as shown below.)

$$2\frac{1}{5}+1\frac{1}{2}$$
$$=2+\frac{1}{5}+1+\frac{1}{2}$$
$$=3+\frac{1}{5}+\frac{1}{2}$$
$$=3+\left(\frac{1}{5}+\frac{1}{2}\right)$$

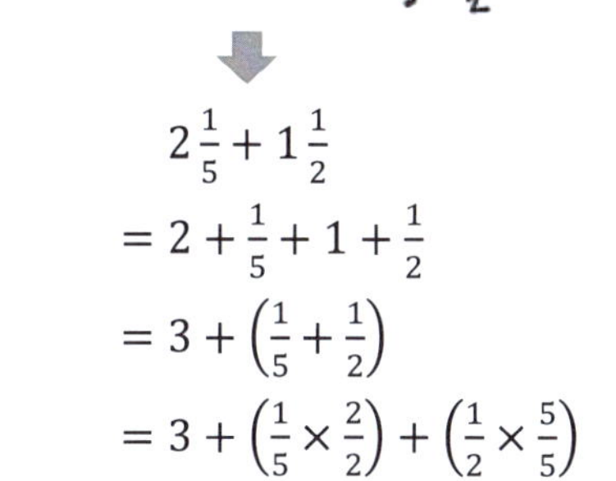

$$2\frac{1}{5}+1\frac{1}{2}$$
$$=2+\frac{1}{5}+1+\frac{1}{2}$$
$$=3+\left(\frac{1}{5}+\frac{1}{2}\right)$$
$$=3+\left(\frac{1}{5}\times\frac{2}{2}\right)+\left(\frac{1}{2}\times\frac{5}{5}\right)$$
$$=3+\frac{2}{10}+\frac{5}{10}$$
$$=3\frac{7}{10}$$

S: (Discuss and find the expression that is true using the commutative and associative properties.)

T: Can we add $\frac{1}{5}+\frac{1}{2}$ without renaming the fractions?

S: No, we need to have like units to add. → We can change fifths and halves to tenths.

T: Yes. We can rename $\frac{1}{2}$ as an equivalent fraction with 10 as the denominator.

T: Say the multiplication sentence for renaming $\frac{1}{5}$ to tenths.

S: $\frac{1}{5}\times\frac{2}{2}=\frac{2}{10}$.

T: Say the multiplication sentence for renaming $\frac{1}{2}$ to tenths.

S: $\frac{1}{2}\times\frac{5}{5}=\frac{5}{10}$.

T: What is our new addition sentence with like units?

S: $3+\frac{2}{10}+\frac{5}{10}=3\frac{7}{10}$.

T: Look at the equations I wrote. (The equations found below the number line.) Discuss with your partner each of the equalities from top to bottom.

S: (Discuss.)

Problem 2: $2\frac{4}{5}+1\frac{1}{2}$

T: Discuss with your partner how expression C is the same as and different from expression B.

S: (Discuss.)

T: Share your thoughts.

S: The sum of the fractional units will be greater than 1 this time.

T: Let's compare them on the number line.

T: (Go through the process quickly, generating each equation. Omit recording the multiplication step for finding equivalent fractions, as shown in the example to the right. Allow 1–2 minutes for students to study these equations.)

$$2\frac{4}{5}+1\frac{1}{2}$$
$$=3\frac{4}{5}+\frac{1}{2}$$
$$=3\frac{8}{10}+\frac{5}{10}$$
$$=3\frac{13}{10}$$
$$=3\frac{10}{10}+\frac{3}{10}$$
$$=4\frac{3}{10}$$

T: If you are ready to find equivalent fractions mentally, do so. If you need to find equivalent fractions by writing the multiplication step, do so.

Problem 3: $2\frac{2}{3} + 5\frac{2}{5}$

$\_\_\_\_ < 2\frac{2}{3} + 5\frac{2}{5} < \_\_\_\_$

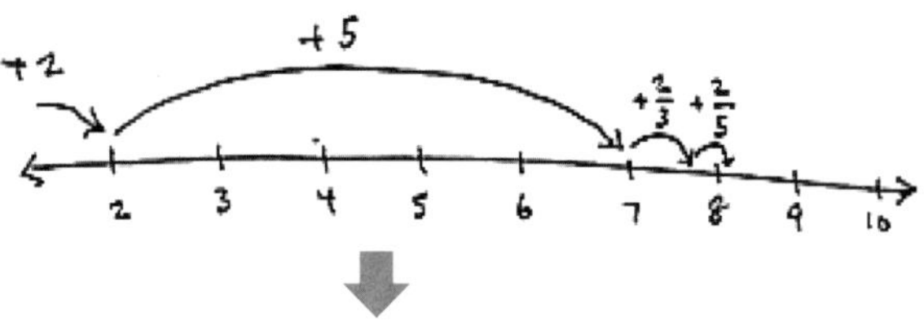

T: (Write the above problem on the board.) The sum will be between which two numbers? Discuss this question with your partner.

S: It's hard to know, because 5 and $\frac{2}{5}$ is really close to 5 and $\frac{2}{6}$. → One way to think about it is that $\frac{2}{6}$ is the same as $\frac{1}{3}$. $\frac{2}{3}$ plus $\frac{1}{3}$ is 1. → 2 + 5 + 1 equals 8, but fifths are larger than sixths. That means the answer must be between 8 and 9 but kind of close to 8.

T: Try solving this problem with your partner. (Post the equations shown to the right.)

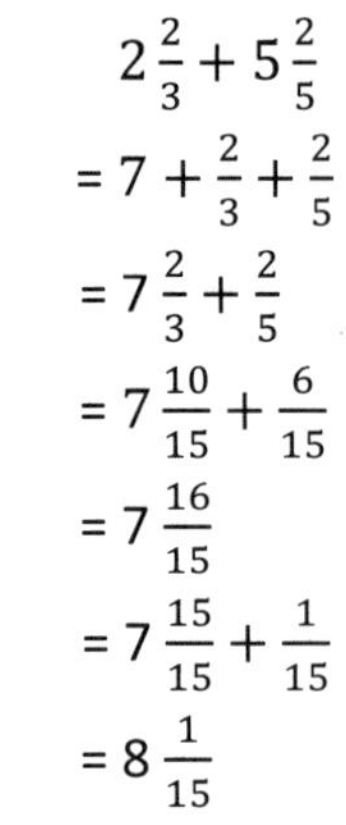

$$2\frac{2}{3} + 5\frac{2}{5}$$
$$= 7 + \frac{2}{3} + \frac{2}{5}$$
$$= 7\frac{2}{3} + \frac{2}{5}$$
$$= 7\frac{10}{15} + \frac{6}{15}$$
$$= 7\frac{16}{15}$$
$$= 7\frac{15}{15} + \frac{1}{15}$$
$$= 8\frac{1}{15}$$

Problem 4: $3\frac{5}{7} + 6\frac{2}{3}$

$\_\_\_\_ < 3\frac{5}{7} + 6\frac{2}{3} < \_\_\_\_$

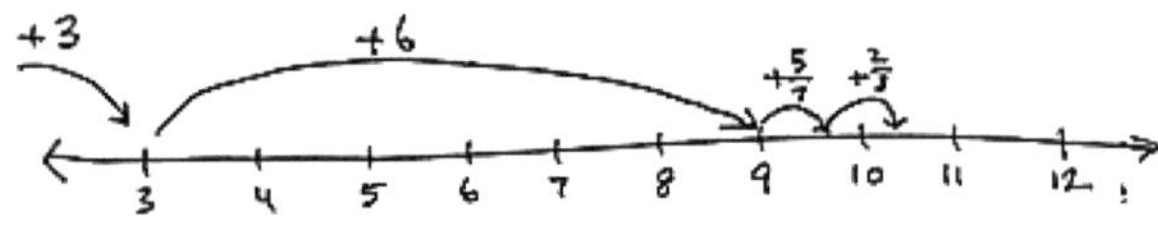

T: (Write the problem on the board.) The sum will be between which two numbers? Discuss this question with your partner.

S: It's greater than 9. → $\frac{5}{7}$ and $\frac{2}{3}$ are both greater than $\frac{1}{2}$, so the answer must be between 10 and 11. → $\frac{5}{7}$ only needs $\frac{2}{7}$ to be 1, and $\frac{2}{3}$ is much more than $\frac{2}{7}$. So, I agree, the answer will be between 10 and 11.

T: Take 2 minutes to solve this problem collaboratively with your partner. (Post the equations shown to the right.)

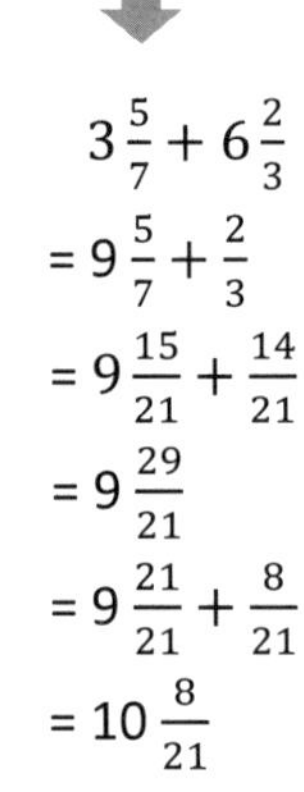

$$3\frac{5}{7} + 6\frac{2}{3}$$
$$= 9\frac{5}{7} + \frac{2}{3}$$
$$= 9\frac{15}{21} + \frac{14}{21}$$
$$= 9\frac{29}{21}$$
$$= 9\frac{21}{21} + \frac{8}{21}$$
$$= 10\frac{8}{21}$$

Problem 5: $3\frac{1}{2} + 4\frac{7}{8}$

$\_\_\_\_ < 3\frac{1}{2} + 4\frac{7}{8} < \_\_\_\_$

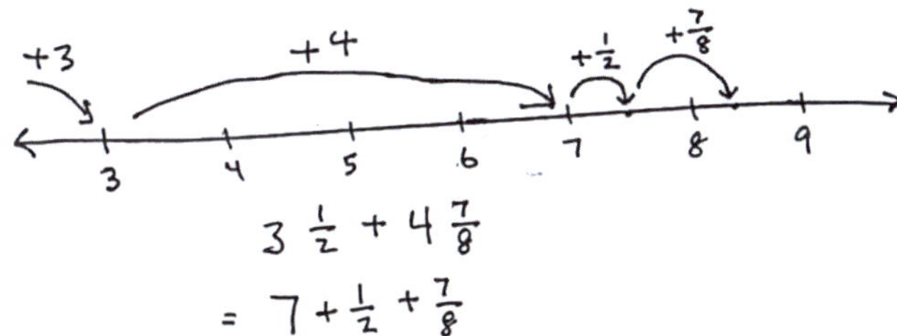

T: (Write the problem on the board.) Discuss with your partner what unit you will use to add the fractional parts. (Allow 1 minute to discuss.)

MP.3

T: Julia and Curtis, you disagree. Julia, what is your choice?

S: I'm just going to use sixteenths. It's easy for me just to multiply the denominators to find a like unit.

T: Curtis, how is your strategy different?

S: Eighths are easier for me because I only have to change the $\frac{1}{2}$ into $\frac{4}{8}$.

T: You have 2 minutes to solve the problem. Rename by using either sixteenths or eighths for like units.

S: (Work.)

Method 1	Method 2
$3\frac{1}{2} + 4\frac{7}{8} =$	$3\frac{1}{2} + 4\frac{7}{8} =$
$7\frac{1}{2} + \frac{7}{8} = 7\frac{4}{8} + \frac{7}{8}$	$7\frac{1}{2} + \frac{7}{8} = 7\frac{8}{16} + \frac{14}{16}$
$= 7\frac{11}{8}$	$= 7\frac{22}{16}$
$= 7\frac{8}{8} + \frac{3}{8}$	$= 7\frac{16}{16} + \frac{6}{16}$
$= 8\frac{3}{8}$	$= 8\frac{6}{16}$
	$= 8\frac{3}{8}$

Allow students two minutes to work together. Though students should strive to simplify their answers, both choices of unit yield an equivalent and correct sum, regardless of the fractions being simplified. Post the equations shown above.

Problem 6: $15\frac{5}{6} + 7\frac{9}{10}$

Allow students to solve the last problem. Again, note that there are two methods for finding like units. As students work, have two pairs come to the board and solve the problems using different units, highlighting that both methods result in the same solution.

It is worth pointing out that, if this were a problem about time, we might want to keep our final fraction as sixtieths in Method 1. The answer might be 23 hours and 44 minutes.

Method 1	Method 2
$15\frac{5}{6} + 7\frac{9}{10}$	$15\frac{5}{6} + 7\frac{9}{10}$
$= 22\frac{5}{6} + \frac{9}{10}$	$= 22\frac{5}{6} + \frac{9}{10}$
$= 22\frac{50}{60} + \frac{54}{60}$	$= 22\frac{25}{30} + \frac{27}{30}$
$= 22\frac{104}{60}$	$= 22\frac{52}{30}$
$= 22\frac{60}{60} + \frac{44}{60}$	$= 22\frac{30}{30} + \frac{22}{30}$
$= 23\frac{11}{15}$	$= 23\frac{11}{15}$

Problem Set (10 minutes)

Students should do their personal best to complete the Problem Set within the allotted 10 minutes. For some classes, it may be appropriate to modify the assignment by specifying which problems they work on first. Some problems do not specify a method for solving. Students should solve these problems using the RDW approach used for Application Problems.

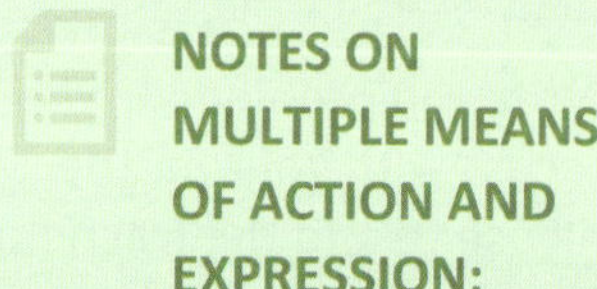

NOTES ON MULTIPLE MEANS OF ACTION AND EXPRESSION:

If students finish early, have them solve the problem using more than one method for finding like units. They might also draw their solutions on the number line to prove the equivalence of different units. Drawings can be shared with the rest of the class to clarify confusion that others may have about the relationship between different methods.

Student Debrief (10 minutes)

Lesson Objective: Add fractions with sums greater than 2.

The Student Debrief is intended to invite reflection and active processing of the total lesson experience.

Invite students to review their solutions for the Problem Set. They should check work by comparing answers with a partner before going over answers as a class. Look for misconceptions or misunderstandings that can be addressed in the Debrief. Guide students in a conversation to debrief the Problem Set and process the lesson.

Any combination of the questions below may be used to lead the discussion.

T: Please take two minutes to check your answers with your partner. Do not change any of your answers. (Allow time for students to work.)

T: I will say the addition problem. Will you please share your answers out loud in response? Problem 1 (a), 2 and 1 fourth + 1 and 1 fifth =...?

S: 3 and 9 twentieths. (Continue with the remainder of the Problem Set.)

T: Take the next two minutes to discuss with your partner any observations you had while completing this Problem Set. What do you notice? (Circulate as students discuss. Listen for conversations that can be shared with the whole class.)

T: Myra, can you share what you noticed happening across the page?

S: Sure. The rows going across shared the same units. Problem 1 (a) and (b) had units of fourths and fifths, and the like units are twentieths.

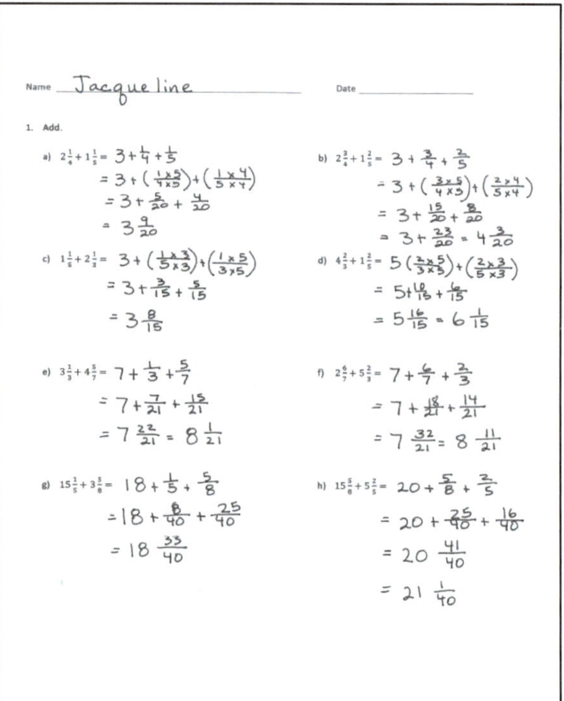

Name Jacqueline Date

1. Add.

a) $2\frac{1}{4}+1\frac{1}{5}= 3+\frac{1}{4}+\frac{1}{5} = 3+\left(\frac{1\times5}{4\times5}\right)+\left(\frac{1\times4}{5\times4}\right) = 3+\frac{5}{20}+\frac{4}{20} = 3\frac{9}{20}$

b) $2\frac{3}{4}+1\frac{2}{5}= 3+\frac{3}{4}+\frac{2}{5} = 3+\left(\frac{3\times5}{4\times5}\right)+\left(\frac{2\times4}{5\times4}\right) = 3+\frac{15}{20}+\frac{8}{20} = 3+\frac{23}{20} = 4\frac{3}{20}$

c) $1\frac{1}{5}+2\frac{1}{3}= 3+\left(\frac{1\times3}{5\times3}\right)+\left(\frac{1\times5}{3\times5}\right) = 3+\frac{3}{15}+\frac{5}{15} = 3\frac{8}{15}$

d) $4\frac{2}{3}+1\frac{2}{5}= 5\left(\frac{2\times5}{3\times5}\right)+\left(\frac{2\times3}{5\times3}\right) = 5+\frac{10}{15}+\frac{6}{15} = 5\frac{16}{15} = 6\frac{1}{15}$

e) $3\frac{1}{3}+4\frac{5}{7}= 7+\frac{1}{3}+\frac{5}{7} = 7+\frac{7}{21}+\frac{15}{21} = 7\frac{22}{21} = 8\frac{1}{21}$

f) $2\frac{6}{7}+5\frac{2}{3}= 7+\frac{6}{7}+\frac{2}{3} = 7+\frac{18}{21}+\frac{14}{21} = 7\frac{32}{21} = 8\frac{11}{21}$

g) $15\frac{1}{5}+3\frac{5}{8}= 18+\frac{1}{5}+\frac{5}{8} = 18+\frac{8}{40}+\frac{25}{40} = 18\frac{33}{40}$

h) $15\frac{5}{8}+5\frac{2}{5}= 20+\frac{5}{8}+\frac{2}{5} = 20+\frac{25}{40}+\frac{16}{40} = 20\frac{41}{40} = 21\frac{1}{40}$

EUREKA MATH™

T: Victor, what did you see in the right column?

S: On all of the problems in the right column, the sum of the fraction was greater than 1. In Problem 1(h) for instance, the answer was 20 and 41 fortieths. 41 fortieths is a fraction greater than 1, so I had to change it into a mixed number and add that to the whole number 20. So, my final answer was 21 and 1 fortieth.

T: Share with your partner how you realize when the fraction allows you to make a new whole. (Allow one minute for conversation.)

S: When the top number of the fraction is greater than the bottom number, I know. → I look at the relationship between the numerator and the denominator. If the numerator is greater, I change it to a mixed number. → The denominator tells us the number of parts in one whole. So, if the numerator is greater, the fraction is greater than one.

T: What about Clayton's reasoning in Problem 4? Discuss your thoughts with your partner.

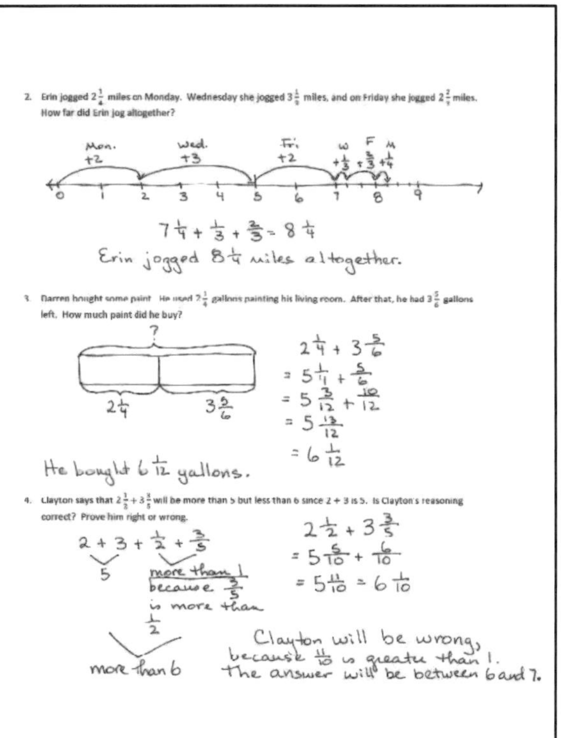

Exit Ticket (3 minutes)

After the Student Debrief, instruct students to complete the Exit Ticket. A review of their work will help with assessing students' understanding of the concepts that were presented in today's lesson and planning more effectively for future lessons. The questions may be read aloud to the students.

A

Number Correct: _______

Add and Subtract Whole Numbers and Ones with Fraction Units

1.	$3 + 1 =$	
2.	$3 + \frac{1}{2} =$	
3.	$3\frac{1}{2} + 1 =$	
4.	$3 - 1 =$	
5.	$3\frac{1}{2} - 1 =$	
6.	$4 - 2 =$	
7.	$4\frac{1}{2} - 2 =$	
8.	$5 - 2 =$	
9.	$5\frac{1}{3} - 2 =$	
10.	$5\frac{2}{3} - 2 =$	
11.	$5\frac{2}{3} + 2 =$	
12.	$6 + 2 =$	
13.	$6 + \frac{3}{4} =$	
14.	$6\frac{3}{4} + 2 =$	
15.	$6\frac{3}{4} - 2 =$	
16.	$6\frac{3}{4} - 3 =$	
17.	$6\frac{3}{4} - 4 =$	
18.	$6\frac{3}{4} - 6 =$	
19.	$6\frac{3}{4} - \frac{3}{4} =$	
20.	$2\frac{5}{6} + 3 =$	
21.	$2\frac{1}{6} + 3 =$	
22.	$2\frac{5}{6} + 7 =$	

23.	$3\frac{5}{6} + 7 =$	
24.	$7\frac{5}{6} + 3 =$	
25.	$10\frac{5}{6} - 3 =$	
26.	$10\frac{5}{6} - 7 =$	
27.	$3 + \frac{4}{5} + 2 =$	
28.	$5 + \frac{7}{8} + 4 =$	
29.	$7 + \frac{4}{5} - 2 =$	
30.	$9 + \frac{5}{12} - 5 =$	
31.	$7 + \frac{1}{5} + \frac{1}{5} + 2 =$	
32.	$7 + \frac{2}{5} + 2 =$	
33.	$7 + \frac{2}{5} + 2 + \frac{2}{5} =$	
34.	$7\frac{2}{5} + 2\frac{2}{5} =$	
35.	$6 + \frac{1}{3} + 1 + \frac{1}{3} =$	
36.	$6\frac{1}{3} + 1\frac{1}{3} =$	
37.	$6 + \frac{2}{3} - 1 =$	
38.	$6\frac{2}{3} - 1\frac{1}{3} =$	
39.	$6\frac{2}{3} - 1\frac{2}{3} =$	
40.	$3 + \frac{4}{7} + 1 + \frac{2}{7} =$	
41.	$3\frac{4}{7} + 1\frac{2}{7} =$	
42.	$7\frac{4}{5} - 2\frac{3}{5} =$	
43.	$7\frac{4}{5} - 2\frac{2}{5} =$	
44.	$13\frac{7}{9} - 7\frac{5}{9} =$	

B

Number Correct: ______

Improvement: ______

Add and Subtract Whole Numbers and Ones with Fraction Units

1.	$2 + 1 =$		23.	$4\frac{5}{6} + 6 =$	
2.	$2 + \frac{1}{2} =$		24.	$6\frac{5}{6} + 4 =$	
3.	$2\frac{1}{2} + 1 =$		25.	$10\frac{5}{6} - 4 =$	
4.	$2 - 1 =$		26.	$10\frac{5}{6} - 6 =$	
5.	$2\frac{1}{2} - 1 =$		27.	$4 + \frac{4}{5} + 2 =$	
6.	$5 - 2 =$		28.	$6 + \frac{7}{8} + 3 =$	
7.	$5\frac{1}{2} - 2 =$		29.	$6 + \frac{4}{5} - 2 =$	
8.	$6 - 2 =$		30.	$9 + \frac{5}{12} - 4 =$	
9.	$6\frac{1}{3} - 2 =$		31.	$6 + \frac{1}{5} + \frac{1}{5} + 2 =$	
10.	$6\frac{2}{3} - 2 =$		32.	$6 + \frac{2}{5} + 2 =$	
11.	$6\frac{2}{3} + 2 =$		33.	$6 + \frac{2}{5} + 2 + \frac{2}{5} =$	
12.	$7 + 2 =$		34.	$6\frac{2}{5} + 2\frac{2}{5} =$	
13.	$7 + \frac{3}{4} =$		35.	$5 + \frac{1}{3} + 1 + \frac{1}{3} =$	
14.	$7\frac{3}{4} + 2 =$		36.	$5\frac{1}{3} + 1\frac{1}{3} =$	
15.	$7\frac{3}{4} - 2 =$		37.	$7 + \frac{2}{3} - 1 =$	
16.	$7\frac{3}{4} - 3 =$		38.	$7\frac{2}{3} - 1\frac{1}{3} =$	
17.	$7\frac{3}{4} - 4 =$		39.	$7\frac{2}{3} - 1\frac{2}{3} =$	
18.	$7\frac{3}{4} - 7 =$		40.	$5 + \frac{4}{7} + 1 + \frac{2}{7} =$	
19.	$7\frac{3}{4} - \frac{3}{4} =$		41.	$5\frac{4}{7} + 1\frac{2}{7} =$	
20.	$3\frac{5}{6} + 2 =$		42.	$6 + \frac{4}{5} - 2\frac{3}{5} =$	
21.	$3\frac{1}{6} + 2 =$		43.	$6\frac{4}{5} - 2\frac{3}{5} =$	
22.	$3\frac{5}{6} + 6 =$		44.	$13\frac{7}{9} - 6\frac{5}{9} =$	

Name ______________________________ Date ______________

1. Add.

a. $2\frac{1}{4} + 1\frac{1}{5} =$

b. $2\frac{3}{4} + 1\frac{2}{5} =$

c. $1\frac{1}{5} + 2\frac{1}{3} =$

d. $4\frac{2}{3} + 1\frac{2}{5} =$

e. $3\frac{1}{3} + 4\frac{5}{7} =$

f. $2\frac{6}{7} + 5\frac{2}{3} =$

EUREKA MATH™

g. $15\frac{1}{5} + 3\frac{5}{8} =$

h. $15\frac{5}{8} + 5\frac{2}{5} =$

2. Erin jogged $2\frac{1}{4}$ miles on Monday. Wednesday, she jogged $3\frac{1}{3}$ miles, and on Friday, she jogged $2\frac{2}{3}$ miles. How far did Erin jog altogether?

3. Darren bought some paint. He used $2\frac{1}{4}$ gallons painting his living room. After that, he had $3\frac{5}{6}$ gallons left. How much paint did he buy?

4. Clayton says that $2\frac{1}{2} + 3\frac{3}{5}$ will be more than 5 but less than 6 since $2 + 3$ is 5. Is Clayton's reasoning correct? Prove him right or wrong.

Name ______________________________ Date ______________

Add.

1. $3\frac{1}{2} + 1\frac{1}{3} =$

2. $4\frac{5}{7} + 3\frac{3}{4} =$

Name ______________________________ Date ______________

1. Add.

a. $2\frac{1}{2} + 1\frac{1}{5} =$

b. $2\frac{1}{2} + 1\frac{3}{5} =$

c. $1\frac{1}{5} + 3\frac{1}{3} =$

d. $3\frac{2}{3} + 1\frac{3}{5} =$

e. $2\frac{1}{3} + 4\frac{4}{7} =$

f. $3\frac{5}{7} + 4\frac{2}{3} =$

g. $15\frac{1}{5} + 4\frac{3}{8} =$

h. $18\frac{3}{8} + 2\frac{2}{5} =$

2. Angela practiced piano for $2\frac{1}{2}$ hours on Friday, $2\frac{1}{3}$ hours on Saturday, and $3\frac{2}{3}$ hours on Sunday. How much time did Angela practice piano during the weekend?

3. String A is $3\frac{5}{6}$ meters long. String B is $2\frac{1}{4}$ meters long. What's the total length of both strings?

4. Matt says that $5 - 1\frac{1}{4}$ will be more than 4, since $5 - 1$ is 4. Draw a picture to prove that Matt is wrong.

Lesson 11

Objective: Subtract fractions making like units numerically.

Suggested Lesson Structure

Fluency Practice	(8 minutes)
Application Problem	(10 minutes)
Concept Development	(32 minutes)
Student Debrief	(10 minutes)
Total Time	**(60 minutes)**

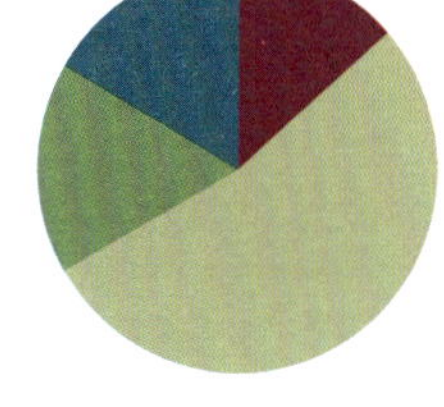

Fluency Practice (8 minutes)

- Subtracting Fractions from Whole Numbers **4.NF.3a** (5 minutes)
- Adding and Subtracting Fractions with Like Units **4.NF.3c** (3 minutes)

Subtracting Fractions from Whole Numbers (5 minutes)

Note: This mental math fluency activity strengthens part–part–whole understanding as it relates to fractions and mixed numbers.

T: I'll say a subtraction sentence. You say the subtraction sentence with the answer. 1 – 1 half.

S: 1 – 1 half = 1 half.

T: 2 – 1 half.

S: 2 – 1 half = 1 and 1 half.

T: 3 – 1 half.

S: 3 – 1 half = 2 and 1 half.

T: 7 – 1 half.

S: 7 – 1 half = 6 and 1 half.

Continue with the following possible sequence:

$1 - \frac{1}{3}$, $1 - \frac{2}{3}$, $2 - \frac{2}{3}$, $2 - \frac{1}{3}$, $5 - \frac{1}{4}$, and $5 - \frac{3}{4}$.

NOTES ON MULTIPLE MEANS OF ACTION AND EXPRESSION:

If students struggle to answer any fluency activity verbally, the teacher can always scaffold the activity by writing what is said. Teachers may consider an alternative that includes drawing on personal white boards, as outlined here:

T: Draw 2 units. (Students draw.)

T: Subtract 1 half. Are we subtracting $\frac{1}{2}$ of 1 unit or both units?

S: Half of 1 unit!

T: Write the number sentence.

S: (Write $2 - \frac{1}{2} = 1\frac{1}{2}$.)

Adding and Subtracting Fractions with Like Units (3 minutes)

Note: This fluency activity reviews adding and subtracting like units mentally.

T: I'll say an addition or subtraction sentence. You say the answer. 3 sevenths + 1 seventh.
S: 4 sevenths.
T: 3 sevenths – 1 seventh.
S: 2 sevenths.
T: 3 sevenths + 3 sevenths.
S: 6 sevenths.
T: 3 sevenths – 3 sevenths.
S: 0.
T: 4 sevenths + 3 sevenths.
S: 1.
T: I'll write an addition sentence. You say true or false. (Write $\frac{2}{5} + \frac{2}{5} = \frac{4}{10}$.)
S: False.
T: Say the sum that makes the addition sentence true.
S: 2 fifths + 2 fifths = 4 fifths.
T: (Write $\frac{5}{8} + \frac{3}{8} = 1$.)
S: True.
T: (Write $\frac{5}{6} + \frac{1}{6} = \frac{6}{12}$.)
S: False.
T: Say the sum that makes the addition sentence true.
S: 5 sixths + 1 sixth = 1.

NOTES ON MULTIPLE MEANS OF ENGAGEMENT:

Assign extension problems to students working above grade level. For example, assign fraction addition and subtraction problems that include the largest fractional unit:

True or false?

$$\frac{3}{4} - \frac{1}{4} = \frac{1}{2}$$

$$\frac{4}{8} + \frac{2}{8} = \frac{3}{4}$$

NOTES ON MULTIPLE MEANS OF ACTION AND EXPRESSION:

The language of whole numbers is much more familiar to English language learners and students working below grade level. Possibly start by presenting the Application Problem with whole numbers.

Meredith went to the movies. She spent $9 of her money on a movie and $8 of her money on popcorn.
How much money did she spend?
If she started with $20, how much is left?

Application Problem (10 minutes)

Meredith went to the movies. She spent $\frac{2}{5}$ of her money on a ticket and $\frac{3}{7}$ of her money on popcorn. How much of her money did she spend? (Extension: How much of her money is left?)

T: Talk with your partner for 30 seconds about strategies to solve this problem. What equation will you use? (Circulate and listen to student responses.)
T: Jackie, will you share?
S: I thought about when I go to the movies and buy a ticket and popcorn. I have to add those two things. So, I am going to add to solve this problem.

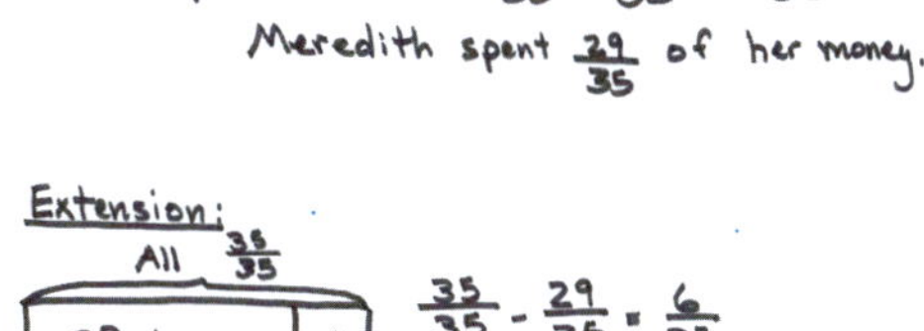

T: Good. David, can you expand on Jackie's comment with your strategy?

S: The units don't match. I need to make like units first, and then I can add the price of the ticket and popcorn together.

T: Nice observation. You have 90 seconds to work with your partner to solve this problem.

S: (Work.)

T: Using the strategies that we learned about adding fractions with unlike units, how can I make like units from fifths and sevenths?

S: Multiply 2 fifths by 7 sevenths, and multiply 3 sevenths by 5 fifths.

T: Everyone, say your addition sentence with your new like units.

S: 14 thirty-fifths plus 15 thirty-fifths equals 29 thirty-fifths.

T: Please share a sentence about the money Meredith spent.

S: Meredith spent 29 thirty-fifths of her money at the theater.

T: Is 29 thirty-fifths more than or less than a whole? How do you know?

S: Less than a whole because the numerator is less than the denominator.

T: (If time allows.) Did anyone answer the extension question?

S: Yes! Her total money would be $\frac{35}{35}$. She spent $\frac{29}{35}$, so $\frac{6}{35}$ is left.

Note: Students solve this Application Problem by making like units numerically to add. This problem also serves as an introduction to this lesson's topic of making like units numerically to subtract.

NOTES ON MULTIPLE MEANS OF REPRESENTATION:

The vignette demonstrated before Problem 1 in the Concept Development uses a conceptual model for finding like units in order to subtract. If students do not need this review, move directly to Problem 1.

Concept Development (32 minutes)

Materials: (S) Personal white board

T: (Write $\frac{1}{3} - \frac{1}{5}$ on the board.) Look at this problem. Tell your partner how you might solve it. (Give 30 seconds for discussion.)

S: I would draw two fraction models. First, I would divide one whole into thirds and bracket $\frac{1}{3}$. Then, I would horizontally divide the other whole into fifths and bracket $\frac{1}{5}$. Then, I would divide both wholes the way the other was divided. That way, I would create like units. Finally, I could subtract.

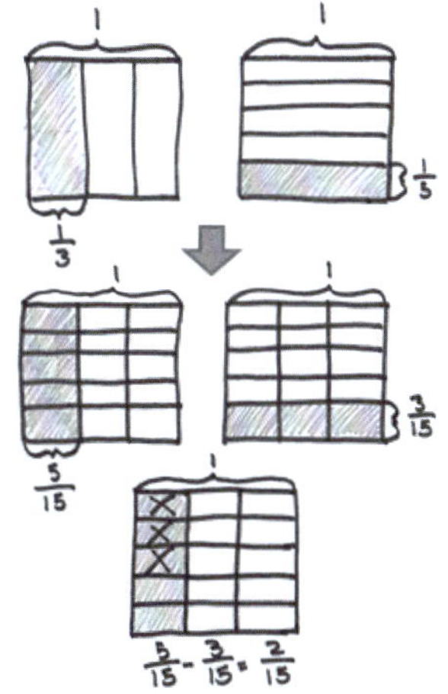

T: What is a like unit for thirds and fifths?

S: Fifteenths.

T: Since we know how to find like units for addition using an equation, let's use that knowledge to subtract using an equation instead of a picture.

Problem 1: $\frac{1}{3} - \frac{1}{5}$

T: How many fifteenths are equal to 1 third?

S: 5 fifteenths.

T: (Write the following on the board.)

$\left(\frac{1}{3} \times \frac{5}{5}\right)$ 5 times as many selected units.
5 times as many units in the whole.

T: How many fifteenths are equal to 1 fifth?

S: 3 fifteenths.

T: (Write the following on the board.)

$\left(\frac{1}{5} \times \frac{3}{3}\right)$ 3 times as many selected units.
3 times as many units in the whole.

T: (Write the following equation on the board.)

$$\left(\frac{1}{3} \times \frac{5}{5}\right) - \left(\frac{1}{5} \times \frac{3}{3}\right) = \frac{5}{15} - \frac{3}{15} =$$

T: As with addition, the equation supports what we drew in our model. Say the subtraction sentence with like units and the answer.

S: 5 fifteenths – 3 fifteenths = 2 fifteenths.

T: (As shown below, write the difference to the subtraction problem.)

$$\frac{5}{15} - \frac{3}{15} = \frac{2}{15}$$

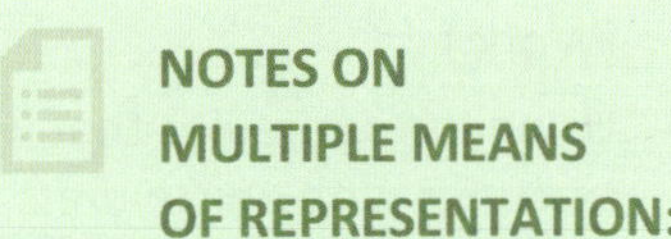

NOTES ON MULTIPLE MEANS OF REPRESENTATION:

Be aware of cognates—words that sound similar and have the same meaning—between English and English language learners' first languages. Related cognates for Spanish speakers, for example, are listed below:

- Fraction = fracción
- Find the sum (add) = sumar
- Numerator = numerador
- Denominator = denominador

Encourage students to listen for cognates and share them with the class. This may help English language learners actively listen, and also boost auditory comprehension as they make links between prior knowledge and new learning.

Problem 2: $\frac{3}{5} - \frac{1}{6}$

T: What do we need to multiply by to make 3 fifths into smaller units?

S: 6 sixths.

T: What do we multiply by to make 1 sixth into smaller units?

S: 5 fifths.

T: (Write the following expression on the board.)

$$\left(\frac{3}{5} \times \frac{6}{6}\right) - \left(\frac{1}{6} \times \frac{5}{5}\right)$$

T: What happened to each fraction?

S: The fractions are still equivalent, but just renamed into smaller units. → We are renaming the fractions into like units so we can subtract them. → We are partitioning our original fractions into smaller units. The value of the fraction doesn't change.

T: Say your subtraction sentence with the like units.

S: 18 thirtieths – 5 thirtieths = 13 thirtieths.

T: (As shown below, write the equation on the board.)

$$\frac{18}{30} - \frac{5}{30} = \frac{13}{30}$$

Problem 3: $1\frac{3}{4} - \frac{3}{5}$

T: What are some different ways we can solve this problem?

S: You can solve it as 2 fifths plus $\frac{3}{4}$. Just take the $\frac{3}{5}$ from 1 to get 2 fifths, and add the 3 fourths. (Shown as Method 1.) → You can subtract the fractional units, and then add the whole number. → I noticed before we started that 3 fifths is less than 3 fourths, so I changed only the fractional units to twentieths. (Shown as Method 2.) → The whole number can be represented as 4 fourths and added to 3 fourths to equal 7 fourths. Then, subtract. (Shown as Method 3.)

Method 1

$1\frac{3}{4} - \frac{3}{5}$

$\frac{5}{5}$ $\frac{3}{4}$

$= \frac{2}{5} + \frac{3}{4}$

$= \frac{8}{20} + \frac{15}{20}$

$= \frac{23}{20}$

$= 1\frac{3}{20}$

Method 2

$1\frac{3}{4} - \frac{3}{5}$

$= 1 + \frac{15}{20} - \frac{12}{20}$

$= 1\frac{3}{20}$

Method 3

$1\frac{3}{4} - \frac{3}{5}$

$\frac{7}{4} - \frac{3}{5}$

$= \frac{35}{20} - \frac{12}{20}$

$= \frac{23}{20}$

$= 1\frac{3}{20}$

Problem 4: $3\frac{3}{5} - 2\frac{1}{2}$

Method 1

$3\frac{3}{5} - 2\frac{1}{2}$

3 $\frac{3}{5}$

$= \frac{1}{2} + \frac{3}{5}$

$= \frac{5}{10} + \frac{6}{10}$

$= 1\frac{1}{10}$

Method 2

$3\frac{3}{5} - 2\frac{1}{2}$

$= 1\frac{3}{5} - \frac{1}{2}$

$= 1 + \frac{6}{10} - \frac{5}{10}$

$= 1\frac{1}{10}$

Method 3

$3\frac{3}{5} - 2\frac{1}{2}$

$= \frac{18}{5} - \frac{5}{2}$

$= \frac{36}{10} - \frac{25}{10}$

$= \frac{11}{10}$

$= 1\frac{1}{10}$

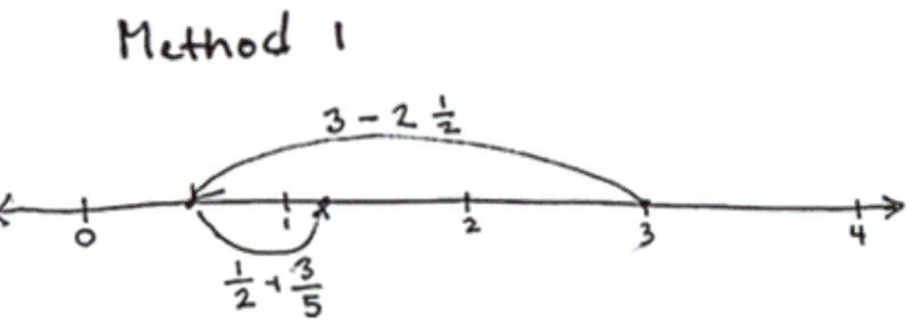

T: (Write the problem on the board.) Solve this problem.

S: (Solve.)

T: (Look for students who solved as above, and have them display their work. Make sure all three methods are represented.)

T: Let's confirm the reasonableness of our answers using the number line to show 2 of our methods.

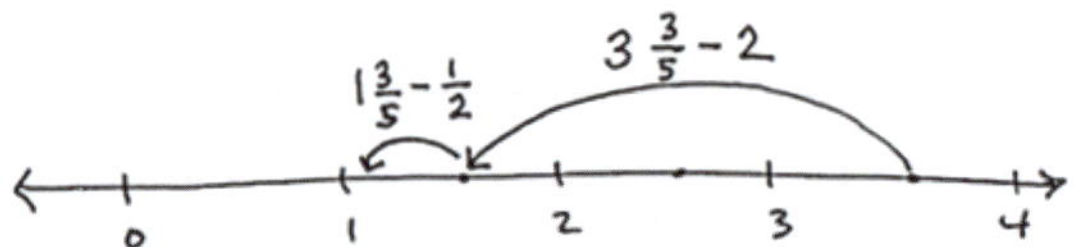

MP.3

T: For Method 1, draw a number line from 0 to 4.

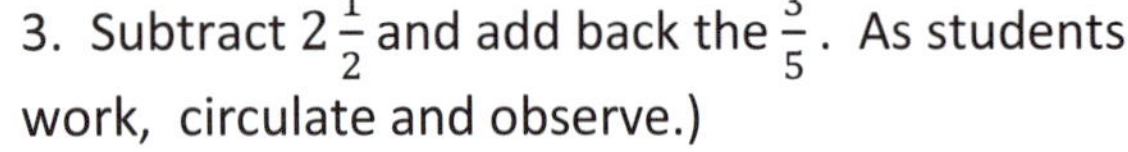

T: (Support students to see that they would start at 3. Subtract $2\frac{1}{2}$ and add back the $\frac{3}{5}$. As students work, circulate and observe.)

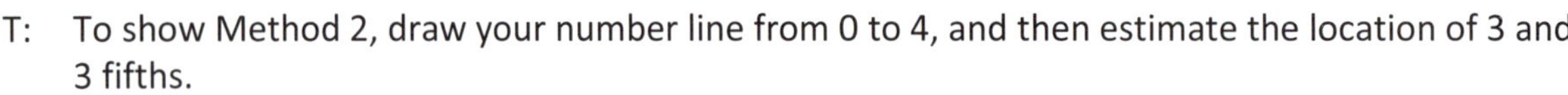

T: To show Method 2, draw your number line from 0 to 4, and then estimate the location of 3 and 3 fifths.

T: Take away 2 first, and then take away the half.

T: Discuss with your partner if the answer of $1\frac{1}{10}$ is reasonable based on both of the number lines.

Problem 5: $5\frac{3}{4} - 3\frac{1}{6}$

T: Estimate the answer first by drawing a number line. The difference between $5\frac{3}{4}$ and $3\frac{1}{6}$ is between which 2 whole numbers?

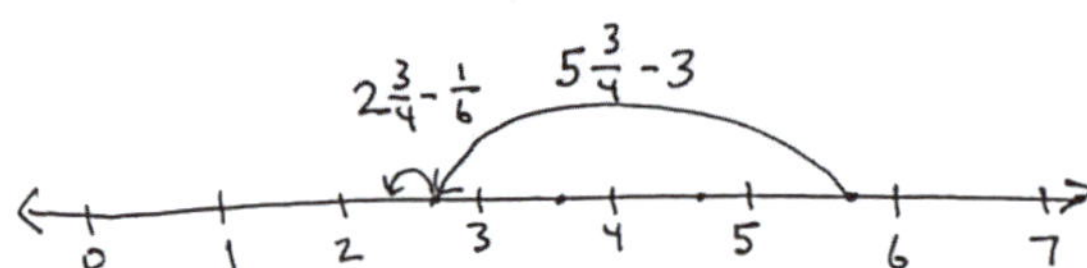

S: Three fourths is much larger than one sixth, so the answer will be between 2 and 3.

T: Will it be closer to 2 or $2\frac{1}{2}$? Discuss your thinking with a partner.

S: (Discuss.)

T: Solve this problem, and find the difference independently. (Circulate and observe as students work.)

S: (Solve.)

T: Some of you used twenty-fourths, and some of you used twelfths to solve this problem. Were your answers the same?

S: They had the same value. $\frac{14}{24}$ can be made into larger units: twelfths. The units are twice as big, so we need half as many to name an equal fraction.

Problem Set (10 minutes)

Students should do their personal best to complete the Problem Set within the allotted 10 minutes. For some classes, it may be appropriate to modify the assignment by specifying which problems they work on first. Some problems do not specify a method for solving. Students should solve these problems using the RDW approach used for Application Problems.

Student Debrief (10 minutes)

Lesson Objective: Subtract fractions making like units numerically.

The Student Debrief is intended to invite reflection and active processing of the total lesson experience.

Invite students to review their solutions for the Problem Set. They should check work by comparing answers with a partner before going over answers as a class. Look for misconceptions or misunderstandings that can be addressed in the Debrief. Guide students in a conversation to debrief the Problem Set and process the lesson.

T: Please take 2 minutes to check your answers with your partner.

T: I will say the subtraction problem. You say your answer out loud. Problem 1(a), 1 half – 1 third.

S: 1 sixth.

T: (Continue with the remaining problems.)

T: Take the next 2 minutes to discuss with your partner any insights you had while solving these problems.

Allow students to discuss, circulating and listening for conversations that can be shared with the whole class.

MP.7

T: Sandy, will you share your thinking about Problem 2?

S: George is wrong. He just learned a rule and thinks it is the only way. It's a good way, but you can also make eighths and sixths into twenty-fourths or ninety-sixths.

T: Discuss in pairs if there are advantages to using twenty-fourths or forty-eighths.

S: Sometimes, it's easier to multiply by the opposite denominator. → Sometimes, larger denominators just get in the way. → Sometimes, they are right. Like if you have to find the minutes, you want to keep your fraction out of 60.

S: An example of this is Problem 1(c). I didn't need to multiply both fractions. I could have just multiplied 3 fourths by 2 halves. Then, I would have had eighths as the like unit for both fractions. Then, the answer is already simplified.

T: Did anyone notice George's issue applying to any of the other problems on the Problem Set?

S: Yes, Problem 1(e). You could use sixtieths or thirtieths. → Yes, in Problem 1(e), the denominator of sixtieths is a small unit, but easy. Thirtieths are a bigger unit and a multiple of both 6 and 10. → Either common unit can be used.

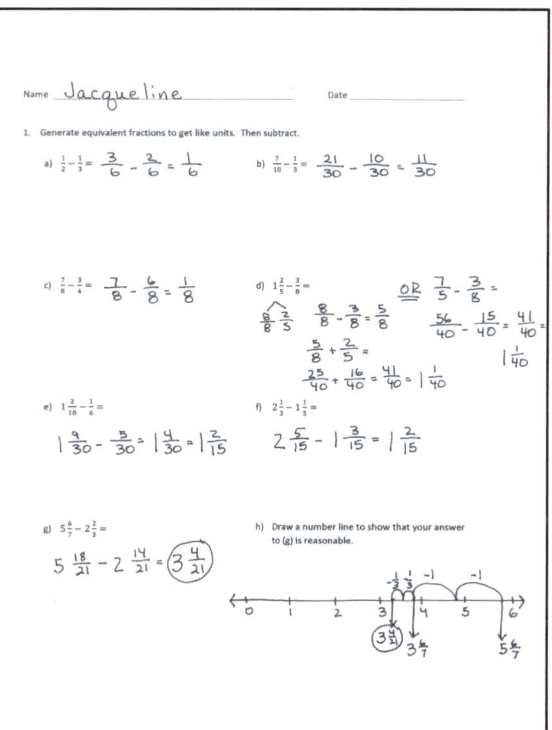

T: I notice that many of you are becoming so comfortable with this equation when subtracting unlike units that you don't have to write the multiplication. You are doing it mentally. However, you still have to check your answers to see if they are reasonable. Discuss with your partner how you use mental math, and also how you make sure your methods and answers are reasonable.

S: It's true. I just look at the other denominator and multiply. It's easy. → I added instead of subtracted and wouldn't have even noticed if I hadn't checked my answer to see that it was greater than the whole amount I started with! → We are learning to find like units, and we may not always need to multiply both fractions. If I don't slow down, I won't even notice there are other choices for solving the problem. → I like choosing the strategy I want to use. Sometimes, it's easier to use the number bond method, and sometimes, it's just easier to subtract from the whole.

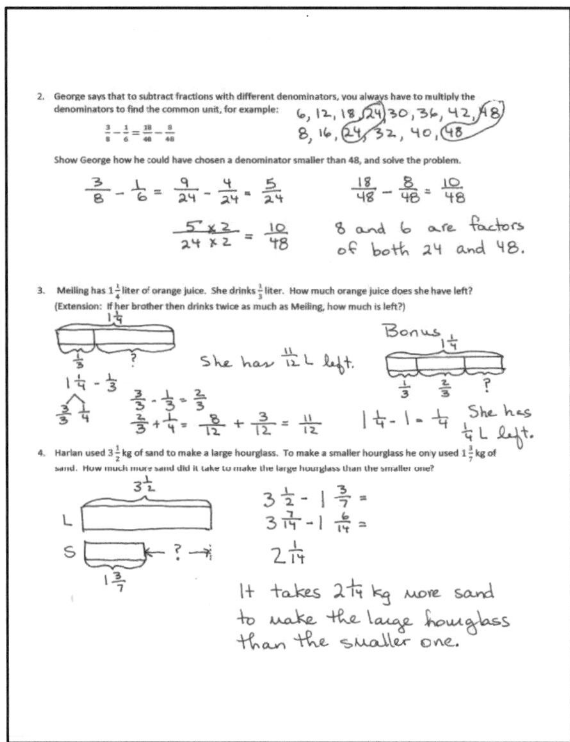
2. George says that to subtract fractions with different denominators, you always have to multiply the denominators to find the common unit, for example:

$\frac{3}{8} - \frac{1}{6} = \frac{18}{48} - \frac{8}{48}$

Show George how he could have chosen a denominator smaller than 48, and solve the problem.

3. Meiling has $1\frac{1}{4}$ liter of orange juice. She drinks $\frac{1}{3}$ liter. How much orange juice does she have left? (Extension: If her brother then drinks twice as much as Meiling, how much is left?)

4. Harlan used $3\frac{1}{2}$ kg of sand to make a large hourglass. To make a smaller hourglass he only used $1\frac{3}{7}$ kg of sand. How much more sand did it take to make the large hourglass than the smaller one?

Exit Ticket (3 minutes)

After the Student Debrief, instruct students to complete the Exit Ticket. A review of their work will help with assessing students' understanding of the concepts that were presented in today's lesson and planning more effectively for future lessons. The questions may be read aloud to the students.

Name ____________________ Date ____________

1. Generate equivalent fractions to get like units. Then, subtract.

a. $\frac{1}{2} - \frac{1}{3} =$

b. $\frac{7}{10} - \frac{1}{3} =$

c. $\frac{7}{8} - \frac{3}{4} =$

d. $1\frac{2}{5} - \frac{3}{8} =$

e. $1\frac{3}{10} - \frac{1}{6} =$

f. $2\frac{1}{3} - 1\frac{1}{5} =$

g. $5\frac{6}{7} - 2\frac{2}{3} =$

h. Draw a number line to show that your answer to (g) is reasonable.

2. George says that, to subtract fractions with different denominators, you always have to multiply the denominators to find the common unit; for example:

$$\frac{3}{8} - \frac{1}{6} = \frac{18}{48} - \frac{8}{48}.$$

Show George how he could have chosen a denominator smaller than 48, and solve the problem.

3. Meiling has $1\frac{1}{4}$ liter of orange juice. She drinks $\frac{1}{3}$ liter. How much orange juice does she have left? (Extension: If her brother then drinks twice as much as Meiling, how much is left?)

4. Harlan used $3\frac{1}{2}$ kg of sand to make a large hourglass. To make a smaller hourglass, he only used $1\frac{3}{7}$ kg of sand. How much more sand did it take to make the large hourglass than the smaller one?

Name ______________________________ Date ______________

Generate equivalent fractions to get like units. Then, subtract.

a. $\frac{3}{4} - \frac{3}{10} =$

b. $3\frac{1}{2} - 1\frac{1}{3} =$

Name ______________________________ Date ______________

1. Generate equivalent fractions to get like units. Then, subtract.

a. $\frac{1}{2} - \frac{1}{5} =$

b. $\frac{7}{8} - \frac{1}{3} =$

c. $\frac{7}{10} - \frac{3}{5} =$

d. $1\frac{5}{6} - \frac{2}{3} =$

e. $2\frac{1}{4} - 1\frac{1}{5} =$

f. $5\frac{6}{7} - 3\frac{2}{3} =$

g. $15\frac{7}{8} - 5\frac{3}{4} =$

h. $15\frac{5}{8} - 3\frac{1}{3} =$

2. Sandy ate $\frac{1}{6}$ of a candy bar. John ate $\frac{3}{4}$ of it. How much more of the candy bar did John eat than Sandy?

3. $4\frac{1}{2}$ yards of cloth are needed to make a woman's dress. $2\frac{2}{7}$ yards of cloth are needed to make a girl's dress. How much more cloth is needed to make a woman's dress than a girl's dress?

4. Bill reads $\frac{1}{5}$ of a book on Monday. He reads $\frac{2}{3}$ of the book on Tuesday. If he finishes reading the book on Wednesday, what fraction of the book did he read on Wednesday?

5. Tank A has a capacity of 9.5 gallons. $6\frac{1}{3}$ gallons of the tank's water are poured out. How many gallons of water are left in the tank?

Lesson 12

Objective: Subtract fractions greater than or equal to 1.

Suggested Lesson Structure

- Fluency Practice (12 minutes)
- Application Problems (10 minutes)
- Concept Development (28 minutes)
- Student Debrief (10 minutes)

Total Time (60 minutes)

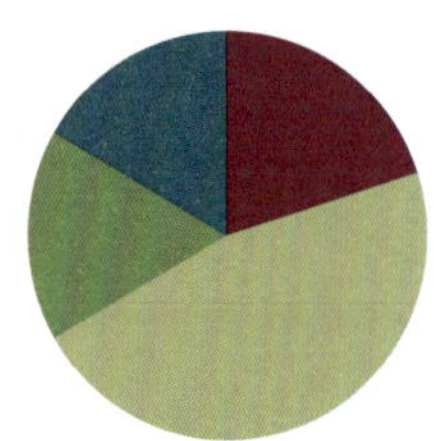

Fluency Practice (12 minutes)

- Sprint: Subtract Fractions with Unlike Units **5.NF.1** (12 minutes)

Sprint: Subtract Fractions with Unlike Units (12 minutes)

Materials: (S) Subtract Fractions with Unlike Units Sprint

Note: This Sprint encourages students to strengthen their mental math strategies while subtracting fractions with unlike units.

NOTES ON MULTIPLE MEANS OF ENGAGEMENT:

There are many ways to differentiate Sprints.

- Allow students to practice Sprints either at home or at school.
- Add a few seconds to Sprint B so students are sure to experience improvement.
- Encourage students who finish all problems before the time is up to list multiples of any given number on the back of their papers.
- Pick Sprints from previous grade levels as appropriate.
- Between Sprint A and Sprint B, allow students to discuss the patterns they observed and used. This may help some students improve their fluency.

Application Problems (10 minutes)

Problem 1

Max's reading assignment was to read $15\frac{1}{2}$ pages. After reading $4\frac{1}{3}$ pages, he took a break. How many more pages does he need to read to finish his assignment?

T: Let's read the problem together.

S: (Read chorally.)

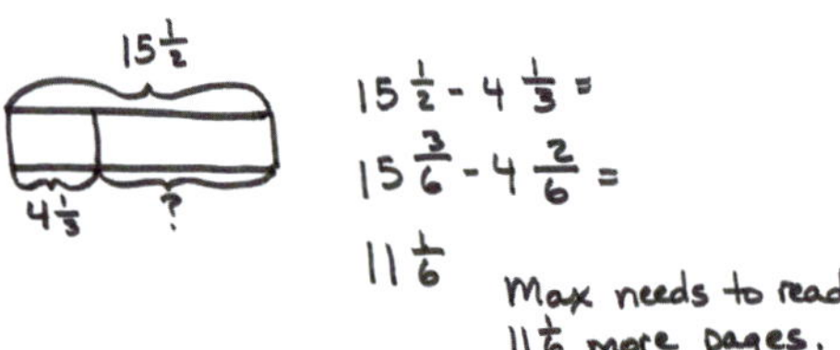

EUREKA MATH™

T: With your partner, share your thoughts about how to solve this problem. (Circulate and listen.)

T: Clara, can you please share your approach?

S: I said that you need to subtract $4\frac{1}{3}$ from $15\frac{1}{2}$ to find the part that is left.

T: Tell me the subtraction problem we need to solve.

S: $15\frac{1}{2} - 4\frac{1}{3}$.

T: Good. This is the same kind of subtraction problem we have been doing since first grade. A part is missing: the pages he has to read to finish.

T: Maggie, read your answer using a complete sentence.

S: Max needs to read 11 and 1 sixth more pages.

NOTES ON MULTIPLE MEANS OF REPRESENTATION:

Problems involving *how many more* are often difficult for students, especially English language learners. Tape diagrams are very helpful as students parse the language of the problem into manageable chunks. Guide students to first determine which tape should be longer, and then label it. Additionally, guide students to use a part–whole model, if necessary.

Problem 2

Sam and Nathan are training for a race. Monday, Sam ran $2\frac{3}{4}$ miles, and Nathan ran $2\frac{1}{3}$ miles. How much farther did Sam run than Nathan?

T: (After students work.) Greg, will you come to the board and show us your solution?

T: (Students can present their solutions for the class to analyze.) Does anyone have questions for Greg?

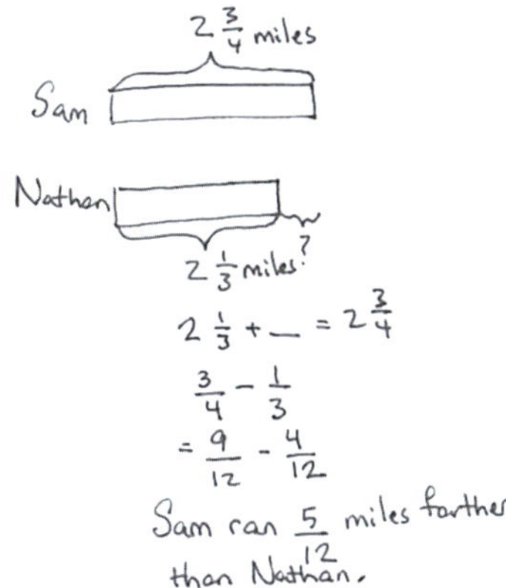

Concept Development (28 minutes)

Materials: (S) Personal white board, empty number line (Lesson 8 Template) or lined paper

Problem 1: $\mathbf{1\frac{1}{2} - \frac{1}{5}}$

$\mathbf{1\frac{1}{5} - \frac{1}{2}}$

T: Look at these 2 problems, and discuss them with your partner.

T: What do you notice?

S: They are the same, except the half and the fifth are switched around.

T: Quickly sketch a number line to show each. Discuss the difference with your partner.

S: (Work.)

T: (Display student work showing the two number lines.)

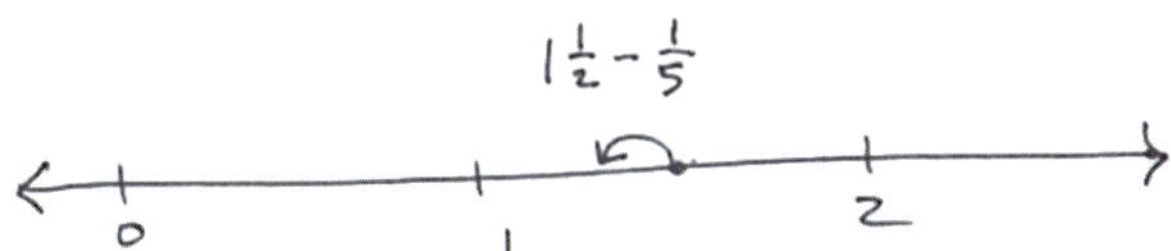

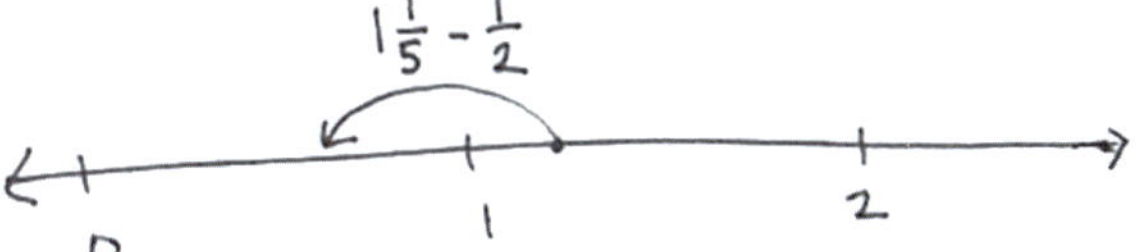

T: You know how to make like units by multiplying. With your partner, show 2 solution methods for $1\frac{1}{5} - \frac{1}{2}$. Show one way taking the half from 1, and the other way taking the half from 1 and 1 fifth.

S: (Work.)

T: (Display both methods, either student or teacher generated.)

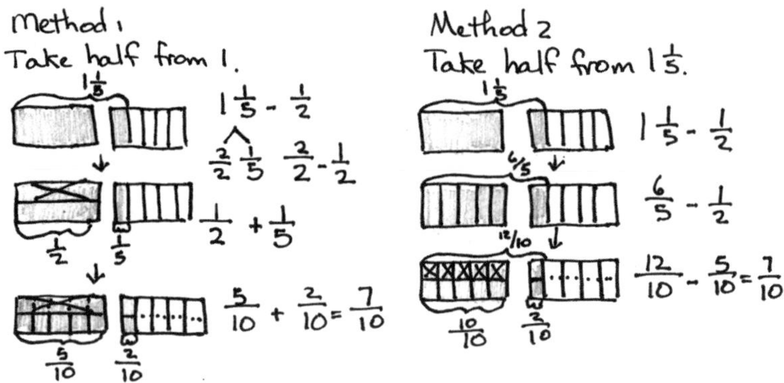

Problem 2: $1\frac{3}{4} - \frac{6}{7}$

T: (Write $1\frac{3}{4} - \frac{6}{7}$ on the board.) Draw a number line. Determine between which two numbers the difference is. Solve this problem using two methods.

S: (Work.)

T: (Display both methods, either student or teacher generated.)

T: Work with your partner to make sure you understand how each step relates to the number lines.

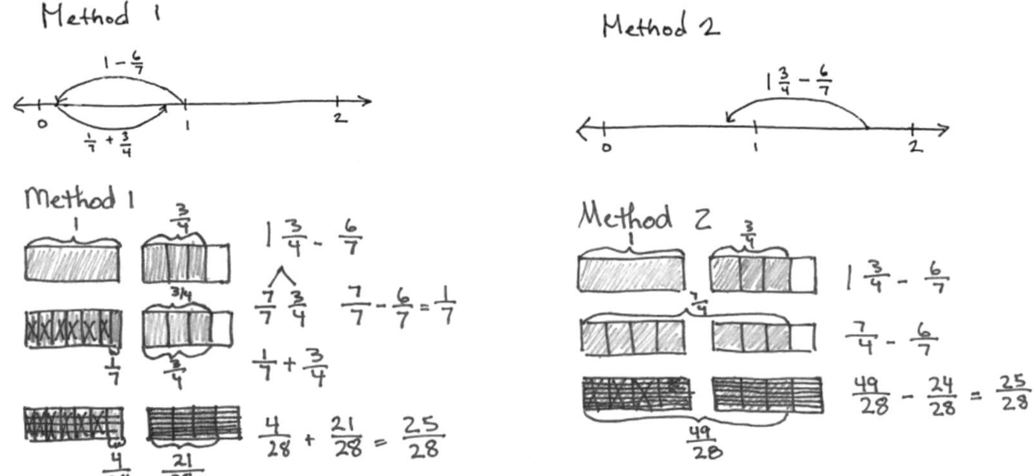

EUREKA MATH™

Problem 3: $3\frac{1}{4} - 2\frac{1}{2}$

T: (Write $3\frac{1}{4} - 2\frac{1}{2}$ on the board.) Draw a number line. Determine between which two numbers the difference is. Solve this problem using two methods.

S: (Work.)

T: (Display both methods, either student or teacher generated.)

T: Work with your partner to make sure you understand how each step relates to the number lines.

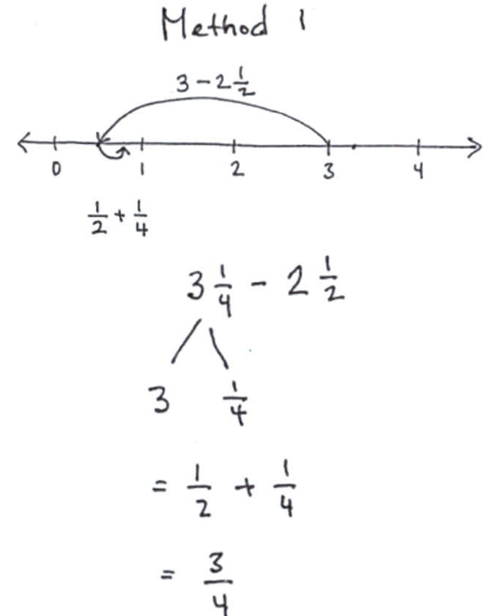

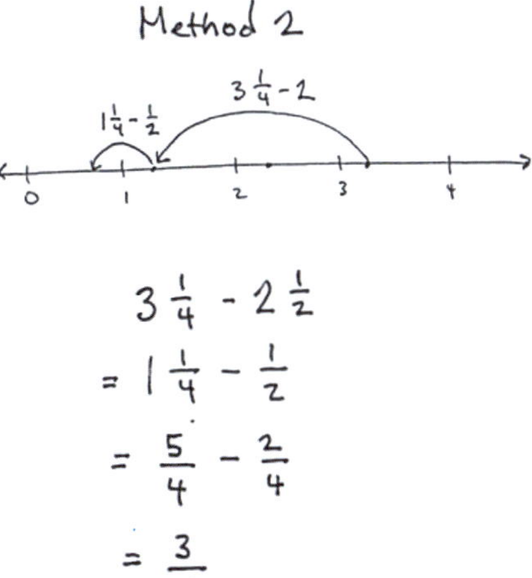

Problem 4: $4\frac{1}{2} - 3\frac{2}{3}$

T: (Write $4\frac{1}{2} - 3\frac{2}{3}$ on the board.) Which is greater, 1 half or 2 thirds?

S: Two thirds.

T: How do you know?

S: Because you just know that 2 thirds is greater than 1 half. → Because if you get like units, you can see that 1 half is the same as 3 sixths, and 2 thirds is the same as 4 sixths.

Students can solve the problem independently or with teacher guidance, depending on their skills and understanding.

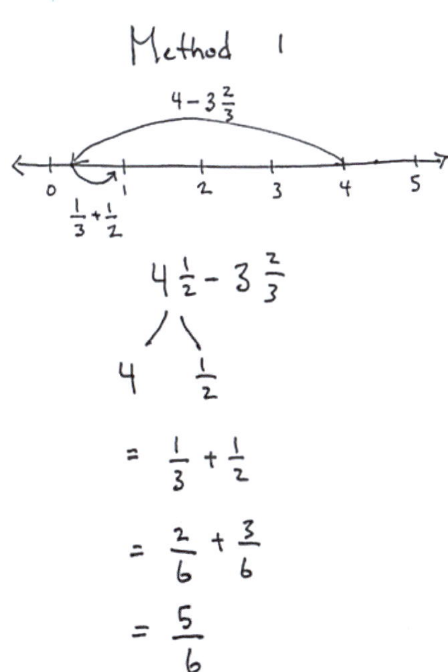

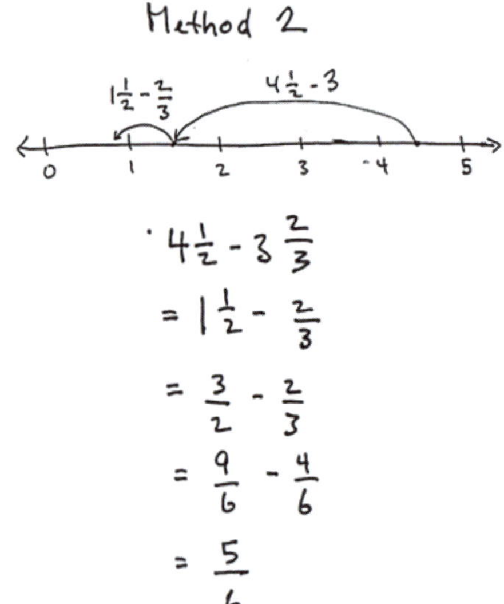

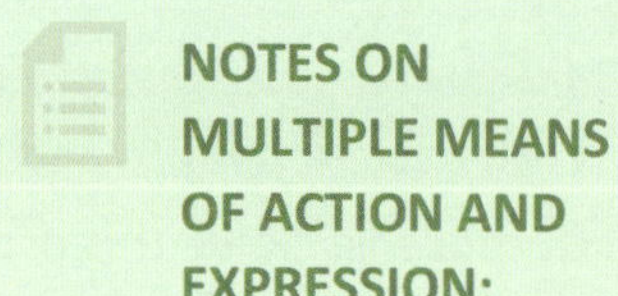

NOTES ON MULTIPLE MEANS OF ACTION AND EXPRESSION:

English language learners may require use of the number line to communicate reasoning about mixed number subtraction long after other students are able to communicate their reasoning without pictorial representations. Continue to allow the option to all students until they independently move away from it.

T: (Post both solution methods on the board.) Analyze the two solution strategies with your partner.

S: Method 1 means taking 3 and 2 thirds from 4 and adding back the half. → Method 2 takes the whole 3 away from $4\frac{1}{2}$ and then subtracts $\frac{2}{3}$.

Problem Set (10 minutes)

Students should do their personal best to complete the Problem Set within the allotted 10 minutes. For some classes, it may be appropriate to modify the assignment by specifying which problems they work on first. Some problems do not specify a method for solving. Students should solve these problems using the RDW approach used for Application Problems.

Student Debrief (10 minutes)

Lesson Objective: Subtract fractions greater than or equal to 1.

The Student Debrief is intended to invite reflection and active processing of the total lesson experience.

Invite students to review their solutions for the Problem Set. They should check work by comparing answers with a partner before going over answers as a class. Look for misconceptions or misunderstandings that can be addressed in the Debrief. Guide students in a conversation to debrief the Problem Set and process the lesson.

T: Take 2 minutes to check your answers with your partner. (Allow students to work.)

T: I will say the problem solutions. Check your work. (Read solutions.)

T: Today, we saw different methods for subtracting. I drew a number bond in some solutions to emphasize how I was thinking about the numbers. Like the number line, it shows a way of thinking. In my work, the number bond shows how I break numbers into parts to make the mathematics easier. Please share with your partner when you used a number bond on your Problem Set.

S: (Share.)

T: Jacqueline, explain why you chose to make a number bond on Problem 1(a).

S: It was clear to me how easy it was to just subtract $2\frac{1}{4}$ from 3. That's just $\frac{3}{4}$. I like adding better anyway, so then I just added the fifth after making like units of twentieths.

T: John, explain why you chose not to make a number bond for 1(g).

S: It just seemed easier to me to subtract the whole numbers first. Right away, I know 17 – 5 is 12.

T: I agree. When I was solving the problems, I also subtracted the whole numbers first on that one for the same reason you gave.

S: I'm noticing that I drew a bond when the numbers were really easy to subtract and their difference was less than 1. I figured out that I would have a friendly fraction to add to the other part.

T: That is precisely the same process you used starting in Grade 1.

MP.7

T: I'm going to list a set of questions. Talk to your partner about how to solve them with a number bond, and how that relates to our work today with fractions.

- Grade 1: 14 – 9.
- Grade 2: 324 – 198.
- Grade 3: 1 foot 3 inches – 7 inches.
- Grade 4: 2 kg – 400 g.
- Grade 5: $1\frac{1}{5} - \frac{3}{7}$.

S: In Grade 1, there weren't enough ones to take from the ones. → In Grade 2, we bonded 324 as 200 and 124, so the answer was just 124 + 2. → In Grade 3, we had to convert 1 foot to 12 inches to take away 7 inches. 12 inches minus 7 is 5 inches plus the 3 extra inches. → In Grade 4, we had to convert 1 kg to 1,000 g to take away 400 g, so we ended up with 1 kg and added back the 600 g. → In Grade 5, we have to convert 1 whole into 7 sevenths to take away 3 sevenths and then add back $\frac{1}{5}$. So, it's 4 sevenths + 1 fifth!

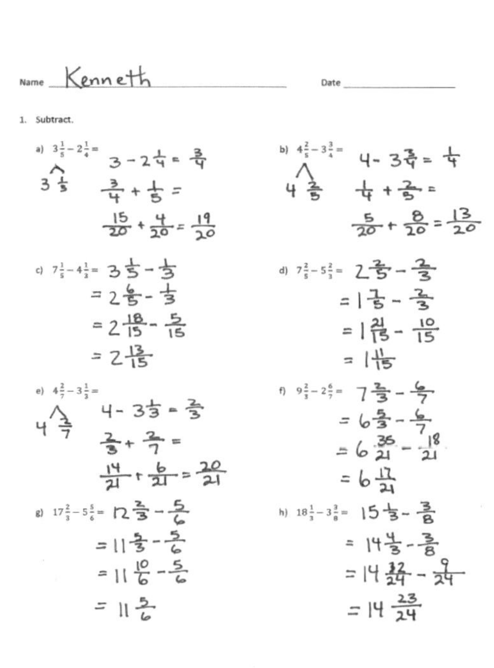
Name Kenneth Date

1. Subtract.

a) $3\frac{1}{5} - 2\frac{1}{4} =$ $3 - 2\frac{1}{4} = \frac{3}{4}$; $\frac{3}{4} + \frac{1}{5} =$; $\frac{15}{20} + \frac{4}{20} = \frac{19}{20}$

b) $4\frac{2}{5} - 3\frac{3}{4} =$ $4 - 3\frac{3}{4} = \frac{1}{4}$; $\frac{1}{4} + \frac{2}{5} =$; $\frac{5}{20} + \frac{8}{20} = \frac{13}{20}$

c) $7\frac{1}{5} - 4\frac{1}{3} = 3\frac{1}{5} - \frac{1}{3}$; $= 2\frac{6}{5} - \frac{1}{3}$; $= 2\frac{18}{15} - \frac{5}{15}$; $= 2\frac{13}{15}$

d) $7\frac{3}{5} - 5\frac{2}{3} = 2\frac{3}{5} - \frac{2}{3}$; $= 1\frac{8}{5} - \frac{2}{3}$; $= 1\frac{24}{15} - \frac{10}{15}$; $= 1\frac{14}{15}$

e) $4\frac{2}{7} - 3\frac{1}{3} =$ $4 - 3\frac{1}{3} = \frac{2}{3}$; $\frac{2}{3} + \frac{2}{7} =$; $\frac{14}{21} + \frac{6}{21} = \frac{20}{21}$

f) $9\frac{2}{3} - 2\frac{6}{7} = 7\frac{2}{3} - \frac{6}{7}$; $= 6\frac{5}{3} - \frac{6}{7}$; $= 6\frac{35}{21} - \frac{18}{21}$; $= 6\frac{17}{21}$

g) $17\frac{2}{3} - 5\frac{5}{6} = 12\frac{2}{3} - \frac{5}{6}$; $= 11\frac{5}{3} - \frac{5}{6}$; $= 11\frac{10}{6} - \frac{5}{6}$; $= 11\frac{5}{6}$

h) $18\frac{1}{3} - 3\frac{3}{8} = 15\frac{1}{3} - \frac{3}{8}$; $= 14\frac{4}{3} - \frac{3}{8}$; $= 14\frac{32}{24} - \frac{9}{24}$; $= 14\frac{23}{24}$

2. Toby wrote the following:

$7\frac{1}{4} - 3\frac{3}{4} = 4\frac{2}{4} = 4\frac{1}{2}$

Is Toby's calculation correct? Draw a number line to support your answer.

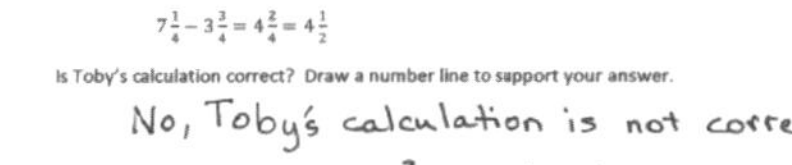
No, Toby's calculation is not correct.

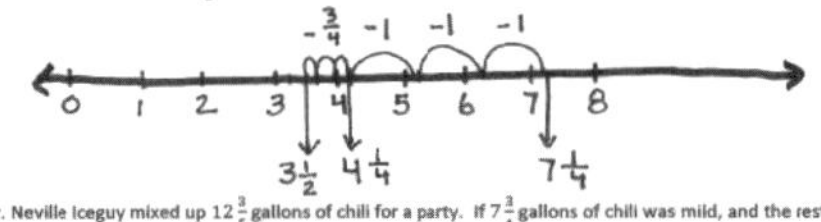

3. Mr. Neville Iceguy mixed up $12\frac{3}{5}$ gallons of chili for a party. If $7\frac{3}{4}$ gallons of chili was mild, and the rest was extra spicy, how much extra spicy chili did Mr. N. Iceguy make?

$12\frac{3}{5} - 7\frac{3}{4}$; $= 5\frac{3}{5} - \frac{3}{4}$; $= 5\frac{12}{20} - \frac{15}{20}$; $= 4\frac{32}{20} - \frac{15}{20} = 4\frac{17}{20}$

Mr. N. Iceguy made $4\frac{17}{20}$ gallons of spicy chili.

4. Jazmyne decided to spend $6\frac{1}{2}$ hours studying over the weekend. She spent $1\frac{1}{4}$ hours studying on Friday evening and $2\frac{2}{3}$ hours on Saturday. How much longer does she need to spend studying on Sunday in order to reach her goal?

$1\frac{1}{4} + 2\frac{2}{3} = 1\frac{3}{12} + 2\frac{8}{12} = 3\frac{11}{12}$

$6\frac{1}{2} - 3\frac{11}{12}$; $= 6\frac{6}{12} - 3\frac{11}{12}$; $= 5\frac{18}{12} - 3\frac{11}{12}$; $= 2\frac{7}{12}$

Jazmyne needs to study $2\frac{7}{12}$ hours on Sunday to reach her goal.

T: Do you notice that every one of the problems has more than one unit? Grade 1 has tens and ones. Grade 2 has hundreds, tens, and ones. Grade 3 has feet and inches. Grade 4 has kilograms and grams, and Grade 5 has whole numbers and fractions. It is important to understand how to play with the units!

Exit Ticket (3 minutes)

After the Student Debrief, instruct students to complete the Exit Ticket. A review of their work will help with assessing students' understanding of the concepts that were presented in today's lesson and planning more effectively for future lessons. The questions may be read aloud to the students.

A

Number Correct: ______

Subtract Fractions with Unlike Units

1.	$^2/_4 - ^1/_4 =$	
2.	$^1/_2 - ^1/_4 =$	
3.	$^2/_6 - ^1/_6 =$	
4.	$^1/_3 - ^1/_6 =$	
5.	$^2/_8 - ^1/_8 =$	
6.	$^1/_4 - ^1/_8 =$	
7.	$^6/_8 - ^1/_8 =$	
8.	$^3/_4 - ^1/_8 =$	
9.	$^3/_4 - ^3/_8 =$	
10.	$^5/_{10} - ^2/_{10} =$	
11.	$^1/_2 - ^2/_{10} =$	
12.	$^1/_2 - ^2/_{10} =$	
13.	$^4/_{10} - ^1/_{10} =$	
14.	$^2/_5 - ^1/_{10} =$	
15.	$^2/_5 - ^3/_{10} =$	
16.	$^6/_{10} - ^3/_{10} =$	
17.	$^3/_5 - ^3/_{10} =$	
18.	$^3/_5 - ^5/_{10} =$	
19.	$^8/_{10} - ^1/_{10} =$	
20.	$^4/_5 - ^1/_{10} =$	
21.	$^4/_5 - ^5/_{10} =$	
22.	$^4/_5 - ^5/_{10} =$	
23.	$^4/_5 - ^7/_{10} =$	
24.	$^2/_{12} - ^1/_{12} =$	
25.	$^1/_6 - ^1/_{12} =$	
26.	$^6/_{12} - ^1/_{12} =$	
27.	$^1/_2 - ^1/_{12} =$	
28.	$^1/_2 - ^5/_{12} =$	
29.	$^{10}/_{12} - ^5/_{12} =$	
30.	$^5/_6 - ^5/_{12} =$	
31.	$^1/_3 - ^3/_{12} =$	
32.	$^2/_3 - ^1/_{12} =$	
33.	$^2/_3 - ^3/_{12} =$	
34.	$^2/_3 - ^7/_{12} =$	
35.	$^1/_4 - ^2/_{12} =$	
36.	$^1/_5 - ^1/_{15} =$	
37.	$^1/_3 - ^1/_{15} =$	
38.	$^2/_3 - ^3/_{15} =$	
39.	$^2/_5 - ^4/_{15} =$	
40.	$^3/_4 - ^2/_{12} =$	
41.	$^3/_4 - ^5/_{16} =$	
42.	$^4/_5 - ^5/_{15} =$	
43.	$^3/_4 - ^4/_{12} =$	
44.	$^3/_4 - ^7/_{16} =$	

B

Number Correct: _______

Improvement: _______

Subtract Fractions with Unlike Units

1.	$^{2}/_{10} - ^{1}/_{10} =$	
2.	$^{1}/_{5} - ^{1}/_{10} =$	
3.	$^{2}/_{4} - ^{1}/_{4} =$	
4.	$^{1}/_{2} - ^{1}/_{4} =$	
5.	$^{5}/_{10} - ^{2}/_{10} =$	
6.	$^{1}/_{2} - ^{2}/_{10} =$	
7.	$^{1}/_{2} - ^{4}/_{10} =$	
8.	$^{4}/_{10} - ^{1}/_{10} =$	
9.	$^{2}/_{5} - ^{1}/_{10} =$	
10.	$^{2}/_{5} - ^{3}/_{10} =$	
11.	$^{6}/_{10} - ^{3}/_{10} =$	
12.	$^{3}/_{5} - ^{3}/_{10} =$	
13.	$^{3}/_{5} - ^{5}/_{10} =$	
14.	$^{8}/_{10} - ^{1}/_{10} =$	
15.	$^{4}/_{5} - ^{1}/_{10} =$	
16.	$^{4}/_{5} - ^{5}/_{10} =$	
17.	$^{4}/_{5} - ^{5}/_{10} =$	
18.	$^{4}/_{5} - ^{7}/_{10} =$	
19.	$^{2}/_{8} - ^{1}/_{8} =$	
20.	$^{1}/_{4} - ^{1}/_{8} =$	
21.	$^{6}/_{8} - ^{1}/_{8} =$	
22.	$^{3}/_{4} - ^{1}/_{8} =$	

23.	$^{3}/_{4} - ^{3}/_{8} =$	
24.	$^{5}/_{15} - ^{1}/_{15} =$	
25.	$^{1}/_{3} - ^{1}/_{15} =$	
26.	$^{3}/_{15} - ^{1}/_{15} =$	
27.	$^{1}/_{5} - ^{1}/_{15} =$	
28.	$^{1}/_{5} - ^{2}/_{15} =$	
29.	$^{12}/_{15} - ^{4}/_{15} =$	
30.	$^{4}/_{5} - ^{4}/_{15} =$	
31.	$^{1}/_{4} - ^{2}/_{12} =$	
32.	$^{3}/_{4} - ^{2}/_{12} =$	
33.	$^{3}/_{4} - ^{4}/_{12} =$	
34.	$^{3}/_{4} - ^{8}/_{12} =$	
35.	$^{1}/_{3} - ^{3}/_{12} =$	
36.	$^{1}/_{6} - ^{1}/_{12} =$	
37.	$^{1}/_{3} - ^{3}/_{15} =$	
38.	$^{2}/_{3} - ^{2}/_{15} =$	
39.	$^{2}/_{5} - ^{2}/_{15} =$	
40.	$^{3}/_{4} - ^{4}/_{12} =$	
41.	$^{3}/_{4} - ^{7}/_{16} =$	
42.	$^{4}/_{5} - ^{4}/_{15} =$	
43.	$^{3}/_{4} - ^{2}/_{12} =$	
44.	$^{3}/_{4} - ^{5}/_{16} =$	

Name ______________________________ Date ________________

1. Subtract.

a. $3\frac{1}{5} - 2\frac{1}{4} =$

b. $4\frac{2}{5} - 3\frac{3}{4} =$

c. $7\frac{1}{5} - 4\frac{1}{3} =$

d. $7\frac{2}{5} - 5\frac{2}{3} =$

e. $4\frac{2}{7} - 3\frac{1}{3} =$

f. $9\frac{2}{3} - 2\frac{6}{7} =$

g. $17\frac{2}{3} - 5\frac{5}{6} =$

h. $18\frac{1}{3} - 3\frac{3}{8} =$

2. Toby wrote the following:

$$7\frac{1}{4} - 3\frac{3}{4} = 4\frac{2}{4} = 4\frac{1}{2}.$$

Is Toby's calculation correct? Draw a number line to support your answer.

3. Mr. Neville Iceguy mixed up $12\frac{3}{5}$ gallons of chili for a party. If $7\frac{3}{4}$ gallons of chili was mild, and the rest was extra spicy, how much extra spicy chili did Mr. Iceguy make?

4. Jazmyne decided to spend $6\frac{1}{2}$ hours studying over the weekend. She spent $1\frac{1}{4}$ hours studying on Friday evening and $2\frac{2}{3}$ hours on Saturday. How much longer does she need to spend studying on Sunday in order to reach her goal?

Name ______________________________ Date ______________

Subtract.

1. $5\frac{1}{2} - 1\frac{1}{3} =$

2. $8\frac{3}{4} - 5\frac{5}{6} =$

Name ______________________________ Date ________________

1. Subtract.

a. $3\frac{1}{4} - 2\frac{1}{3} =$

b. $3\frac{2}{3} - 2\frac{3}{4} =$

c. $6\frac{1}{5} - 4\frac{1}{4} =$

d. $6\frac{3}{5} - 4\frac{3}{4} =$

e. $5\frac{2}{7} - 4\frac{1}{3} =$

f. $8\frac{2}{3} - 3\frac{5}{7} =$

g. $18\frac{3}{4} - 5\frac{7}{8} =$

h. $17\frac{1}{5} - 2\frac{5}{8} =$

2. Tony wrote the following:

$$7\frac{1}{4} - 3\frac{3}{4} = 4\frac{1}{4} - \frac{3}{4}.$$

Is Tony's statement correct? Draw a number line to support your answer.

3. Ms. Sanger blended $8\frac{3}{4}$ gallons of iced tea with some lemonade for a picnic. If there were $13\frac{2}{5}$ gallons of the beverage, how many gallons of lemonade did she use?

4. A carpenter has $10\frac{1}{2}$ feet of wooden plank. He cuts off $4\frac{1}{4}$ feet to replace the slat of a deck and $3\frac{2}{3}$ feet to repair a bannister. He uses the rest of the plank to fix a stair. How many feet of wood does the carpenter use to fix the stair?

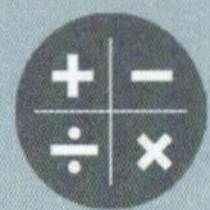

Topic D

Further Applications

5.NF.1, 5.NF.2

Focus Standards:		5.NF.1	Add and subtract fractions with unlike denominators (including mixed numbers) by replacing given fractions with equivalent fractions in such a way as to produce an equivalent sum or difference of fractions with like denominators. *For example, 2/3 + 5/4 = 8/12 + 15/12 = 23/12. (In general, a/b + c/d = (ad + bc)/bd.)*
		5.NF.2	Solve word problems involving addition and subtraction of fractions referring to the same whole, including cases of unlike denominators, e.g., by using visual fraction models or equations to represent the problem. Use benchmark fractions and number sense of fractions to estimate mentally and assess the reasonableness of answers. *For example, recognize an incorrect result 2/5 + 1/2 = 3/7, by observing that 3/7 < 1/2.*
Instructional Days:		4	
Coherence	**-Links from:**	G4–M5	Fraction Equivalence, Ordering, and Operations
	-Links to:	G5–M1	Place Value and Decimal Fractions
		G5–M4	Multiplication and Division of Fractions and Decimal Fractions

Topic D opens with students estimating the value of expressions involving sums and differences with fractions. "Will your sum be less than or greater than one half? One? How do you know?" Though these conversations have been embedded within almost every Student Debrief up to this point, by setting aside an instructional day to dig deeply into logical arguments, students see that it is very easy to forget to make sense of numbers when calculating. This is really the theme of this topic—reasoning while using fractions.

In Lesson 14, students look for number relationships before calculating, for example, to use the associative property or part–whole understanding. Looking for relationships allows them to see shortcuts and connections that are so often bypassed in the rush to get the answer.

In Lesson 15, students solve multi-step word problems and actively assess the reasonableness of their answers. In Lesson 16, they explore part–whole relationships while solving a challenging problem: "One half of Nell's money is equal to 2 thirds of Jennifer's." This lesson challenges the underlying assumption of all fraction arithmetic—that when adding and subtracting, fractions are always defined in relationship to the same whole amount. The beauty of this exploration is to see students grasp that $\frac{1}{2}$ of one thing can be equivalent to $\frac{2}{3}$ of another!

A Teaching Sequence Toward Mastery of Further Applications	
Objective 1:	**Use fraction benchmark numbers to assess reasonableness of addition and subtraction equations.** **(Lesson 13)**
Objective 2:	**Strategize to solve multi-term problems.** **(Lesson 14)**
Objective 3:	**Solve multi-step word problems; assess reasonableness of solutions using benchmark numbers.** **(Lesson 15)**
Objective 4:	**Explore part-to-whole relationships.** **(Lesson 16)**

Lesson 13

Objective: Use fraction benchmark numbers to assess reasonableness of addition and subtraction equations.

Suggested Lesson Structure

Fluency Practice	(11 minutes)
Application Problem	(7 minutes)
Concept Development	(32 minutes)
Student Debrief	(10 minutes)
Total Time	**(60 minutes)**

Fluency Practice (11 minutes)

- From Fractions to Decimals **4.NF.6** (5 minutes)
- Adding and Subtracting Fractions with Unlike Units **5.NF.1** (6 minutes)

From Fractions to Decimals (5 minutes)

Note: This fluency activity reviews decimals as they relate to generating equivalent benchmark fractions.

T: (Write $\frac{1}{10}$.) Say the fraction in unit form.
S: 1 tenth.
T: Say the fraction in decimal form.
S: Zero point one.
T: I'll say a fraction in unit form. You say the fraction in decimal form. Ready? 3 tenths.
S: Zero point three.
T: 7 tenths.
S: Zero point seven.
T: (Write $\frac{1}{2} = \frac{}{10}$.) Say the equivalent fraction with the missing numerator.
S: 1 half = 5 tenths.
T: Say 5 tenths as a decimal.
S: Zero point five.
T: Say 1 half as a decimal.
S: Zero point five.

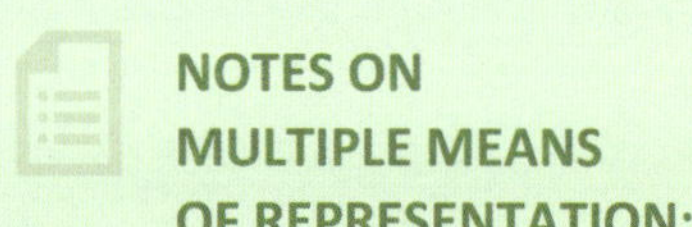

NOTES ON MULTIPLE MEANS OF REPRESENTATION:

If students don't remember how to convert from fractions to decimals, then consider doing a review with the whole class. Fractions can be converted to decimals easily when the denominator is tenths, hundredths, or thousandths. It's just like converting the fractions into equivalent fractions with the denominator of tenths, hundredths, or thousandths.

The teacher can also draw out the fraction bars that clearly show equivalent fractions. ($\frac{2}{5} = \frac{4}{10} = 0.4$. Both $\frac{2}{5}$ and $\frac{4}{10}$ have the same values.) The bar models serve as a great visual.

T: Say 3 and 1 half as a decimal.

S: Three point five.

Continue with the following possible sequence: $\frac{1}{5}$, $\frac{2}{5}$, $\frac{4}{5}$, $2\frac{4}{5}$, $\frac{1}{4}$, $\frac{3}{4}$, $5\frac{3}{4}$, $\frac{1}{25}$, $\frac{2}{25}$, $\frac{3}{25}$, $5\frac{3}{25}$, $\frac{1}{20}$, $\frac{11}{20}$, $3\frac{11}{20}$, $\frac{1}{50}$, $\frac{3}{50}$, and $4\frac{3}{50}$.

Adding and Subtracting Fractions with Unlike Units (6 minutes)

Materials: (S) Personal white board

Note: Students review adding unlike units and practice assessing the reasonableness of a sum in preparation for today's Concept Development.

T: (Write $\frac{1}{4} + \frac{1}{2} = \frac{2}{6}$.) True or false?

S: False.

T: On your personal white board, write the answer that makes the addition sentence true.

S: (Write $\frac{1}{4} + \frac{1}{2} = \frac{1}{4} + \frac{2}{4} = \frac{3}{4}$.)

T: (Write $\frac{1}{2} + \frac{3}{8} = \frac{7}{8}$.) True or false?

S: True.

T: Rewrite the addition sentence using like units.

S: (Write $\frac{1}{2} + \frac{3}{8} = \frac{4}{8} + \frac{3}{8} = \frac{7}{8}$.)

T: (Write $\frac{2}{3} - \frac{2}{9} = \frac{4}{9}$.) True or false?

S: True.

T: Rewrite the subtraction sentence using like units.

S: (Write $\frac{2}{3} - \frac{2}{9} = \frac{6}{9} - \frac{2}{9} = \frac{4}{9}$.)

T: (Write $\frac{5}{6} - \frac{2}{3} = \frac{3}{3}$.) True or false?

S: False.

T: Write the answer that will make the subtraction sentence true.

S: (Write $\frac{5}{6} - \frac{2}{3} = \frac{5}{6} - \frac{4}{6} = \frac{1}{6}$.)

NOTES ON MULTIPLE MEANS OF ENGAGEMENT:

Provide "thinking time" for students to process the problem before answering true or false. If necessary, consider giving another few seconds for students to discuss the problem with their partners. Perhaps have them explain to their partners why they think a problem is true or false.

Application Problem (7 minutes)

Mark jogged $3\frac{5}{7}$ km. His sister jogged $2\frac{4}{5}$ km. How much farther did Mark jog than his sister?

Remind students to approach the problem with the RDW strategy. This is a very brief Application Problem. While circulating as students work, quickly assess which work to select for a short two- or three-minute Debrief.

Note: Students solve this Application Problem involving addition and subtraction of fractions greater than 2 and having unlike denominators, using visual models.

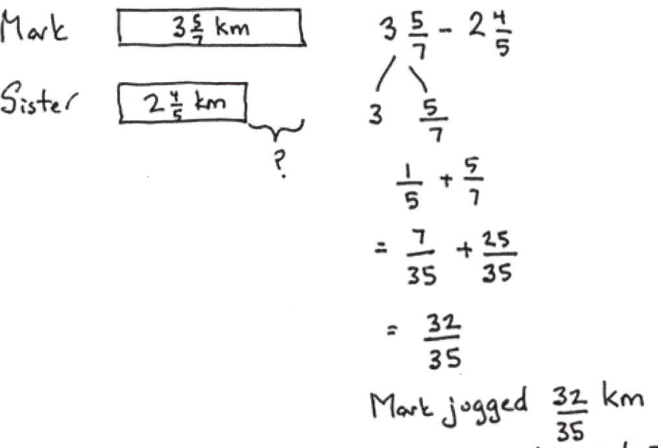

NOTES ON MULTIPLE MEANS OF REPRESENTATION:

If students are not ready to estimate the sum or difference of a fraction sentence, consider doing a mini pre-lesson or a fluency activity on estimating a single fraction before moving on to this lesson.

For example:

- Is $\frac{3}{4}$ closer to 0, $\frac{1}{2}$, or 1 whole?
- Is $\frac{2}{7}$ closer to 0, $\frac{1}{2}$, or 1 whole?
- Is $\frac{9}{10}$ closer to 0, $\frac{1}{2}$, or 1 whole?
- Is $1\frac{6}{7}$ closer to 1, $1\frac{1}{2}$, or 2?
- Is $3\frac{4}{7}$ closer to 3, $3\frac{1}{2}$, or 4?

If necessary, write each fraction on a sentence strip (using it as a number line) and label it. This way, students can easily see whether the fraction is closer to 0, $\frac{1}{2}$, or 1 whole.

Concept Development (32 minutes)

Materials: (S) Personal white board

Problem 1: $\frac{1}{2}+\frac{3}{4}$

T: For the past two weeks we have been learning different strategies to add and subtract unlike fractions. Today the focus is on mental math—using reasoning without actually solving using paper and pencil.

T: (Write $\frac{1}{2}+\frac{3}{4}$.) Think about this expression without solving it using paper and pencil. Share your analysis with a partner.

S: $\frac{1}{2}$ could be 50% of something or 50 cents of a dollar. → I know that $\frac{1}{2}$ is the same as $\frac{5}{10}$ or 0.5 as a decimal. → I know $\frac{3}{4}$ is more than half because half of a whole is $\frac{2}{4}$. → $\frac{3}{4}$ is the same as 75%; 3 quarters equal 75 cents.

T: What do you know about the total value of this expression without solving?

MP.3

S: Since $\frac{3}{4}$ is more than half and we need to add $\frac{1}{2}$ more, the answer will be greater than 1. → $\frac{1}{2}+\frac{3}{4}>1$. → It's like adding 50 cents and 75 cents. The answer will be more than 1 dollar. → $\frac{1}{2}+\frac{2}{4}=1$, but there's still a $\frac{1}{4}$ to add. → The total answer is $1\frac{1}{4}$.

NOTES ON MULTIPLE MEANS OF ENGAGEMENT:

When students are solving problems with partners and continue to struggle even with guided questions, consider asking them to use personal white boards to draw number lines. They can estimate one fraction at a time and then estimate the final answer.

The following is an example:

$\frac{4}{10}+\frac{1}{3}$.

Draw a number line for $\frac{4}{10}$. $\frac{4}{10}$ is less than 1 half.

Draw a number line for $\frac{1}{3}$. $\frac{1}{3}$ is less than 1 half.

Less than $\frac{1}{2}$ + less than $\frac{1}{2}<1$.

$\frac{4}{10}+\frac{1}{3}<1$.

Problem 2: $1\frac{2}{5}-\frac{2}{3}$

T: (Write $1\frac{2}{5}-\frac{2}{3}$.) Without calculating, what do you know about the value of this expression? Talk to your partner.

S: I see that it's a subtraction problem. $\frac{2}{5}$ is less than $\frac{1}{2}$, and $\frac{2}{3}$ is more than $\frac{1}{2}$. → I know that $\frac{2}{3}$ can't be subtracted from $\frac{2}{5}$ because $\frac{2}{3}$ is larger, so we'll need to subtract from 1 whole. → I can convert $1\frac{2}{5}$ to $\frac{7}{5}$ in my head.

T: Do you think the answer is more than 1 or less than 1? Turn and share.

S: Less than 1 because $\frac{1}{5}$ is less than $\frac{1}{3}$, so $\frac{2}{5}$ is less than $\frac{2}{3}$. → The answer is less than 1 because I can create equivalent fractions in my head and solve.

Problem 3: $\frac{4}{10}+\frac{1}{3}$

T: (Write $\frac{4}{10}+\frac{1}{3}$.) Use reasoning skills to decide if the sum is more than or less than 1. Work with your partner.

NOTES ON MULTIPLE MEANS OF ENGAGEMENT:

Because both addends are clearly less than half, this is an easy question. For students working above grade level, let them determine if $\frac{3}{10}+\frac{2}{3}$ is less than, equal to, or greater than 1. Encourage them *not* to solve the problem until they have determined their reasoning.

Allow a minute for students to analyze and discuss the problem. Circulate and listen. If students seem to be lost or off track with their thinking, then perhaps use some of the following questions:

- Is $\frac{4}{10}$ more than 1 half or less than 1 half?
- Is $\frac{4}{10}$ closer to 0 or 1 whole?
- What's half of 10 tenths?
- What's $\frac{4}{10}$ as a decimal?
- How much money is 4 tenths?
- Is $\frac{1}{3}$ closer to 0 or 1 whole?
- Is $\frac{1}{3}$ more than 1 half or less than 1 half?

NOTES ON MULTIPLE MEANS OF ENGAGEMENT:

Have students working below grade level try $\frac{4}{10} + \frac{1}{9}$. Have students working above grade level try $\frac{3}{10} + \frac{3}{9}$.

Problem 4: $\frac{4}{10} + \frac{2}{9}$

MP.3

T: (Write $\frac{4}{10} + \frac{2}{9}$.) Share your analysis of this expression with your partner.

S: I see that it's an addition problem. $\frac{4}{10}$ is less than $\frac{1}{2}$ because $\frac{4}{10} = 0.4$. → I agree. I also noticed that $\frac{2}{9}$ is less than $\frac{1}{2}$ because $\frac{1}{2}$ of 9 is $4\frac{1}{2}$, and 2 is less than $4\frac{1}{2}$. → Both fractions are closer to 0 than 1.

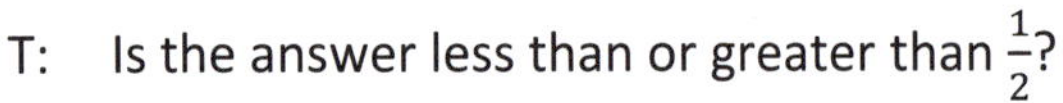

NOTES ON MULTIPLE MEANS OF ENGAGEMENT:

Have students working below grade level try $1\frac{4}{7} - 1$. Have students working above grade level try $1\frac{3}{7} - \frac{9}{10}$.

T: Is the answer less than or greater than $\frac{1}{2}$?

S: $\frac{4}{10}$ is really close to a half. It only needs $\frac{1}{10}$ to be one half. → I'm asking myself, *is* $\frac{2}{9}$ *greater than* $\frac{1}{10}$? If it is, the answer will be greater than $\frac{1}{2}$. → $\frac{2}{9}$ has to be greater than $\frac{1}{10}$ because it's close to $\frac{1}{4}$ or $\frac{2}{8}$.

T: Verify your answer.

Problem 5: $1\frac{4}{7} - \frac{9}{10}$

T: (Write $1\frac{4}{7} - \frac{9}{10}$.) Think about this expression with your partner.

S: $\frac{4}{7}$ is more than $\frac{1}{2}$, and $\frac{9}{10}$ is $\frac{1}{10}$ away from 1 whole. → I know that $\frac{9}{10}$ can't be subtracted from $\frac{4}{7}$, because $\frac{9}{10}$ is larger, so we'll need to subtract from 1 whole. → I would use $1 - \frac{9}{10} = \frac{1}{10}$. → I agree. Now, we have a leftover of $\frac{1}{10} + \frac{4}{7}$.

T: Is the value of this expression greater than or less than $\frac{1}{2}$?

S: I think the value is more than $\frac{1}{2}$ because I know $\frac{4}{7}$ alone is already more than $\frac{1}{2}$. → 1 less than $1\frac{4}{7}$ is going to be more than half. $\frac{9}{10}$ is less than 1, so $\frac{9}{10}$ less than $1\frac{4}{7}$ is going to be greater than $\frac{1}{2}$.

Problem 6: $\frac{4}{5} - \frac{1}{8}$

T: (Write $\frac{4}{5} - \frac{1}{8}$.) Discuss this problem with your partner. Is the value of the expression more than or less than $\frac{1}{2}$?

Use the following questions for support:

- Is $\frac{4}{5}$ more than 1 half or less than 1 half?
- Is $\frac{4}{5}$ closer to 0 or 1 whole?
- What's half of 5 fifths?
- Can you convert 4 fifths to tenths or a decimal in your head? What is it?
- Is $\frac{1}{8}$ more than 1 half or less than 1 half?
- Is $\frac{1}{8}$ closer to 0 or 1 whole?

Problem 7: $2\frac{1}{3} + 3\frac{1}{5}$ ____ $6 + \frac{7}{8}$

T: (Write $2\frac{1}{3} + 3\frac{1}{5}$ ____ $6 + \frac{7}{8}$.) Which expression is greater? Share your thinking with your neighbor.

S: I first need to estimate the total for both expressions, and then I can compare them. → I'll first add up the whole numbers on the left, and then compare them because they're the larger place values. If the whole numbers are equal, then I'll estimate the fractions and compare them.

Allow two minutes for students to analyze and discuss the problem. Circulate and listen. If students seem to be lost or off-track with their thinking, use the following questions to guide discussion and thinking:

- What do you know about $2\frac{1}{3}$ and $3\frac{1}{5}$?
- What's the total of the whole numbers on the left?
- How do you compare the whole numbers?
- What do you know about $\frac{1}{3}$ and $\frac{1}{5}$?
- Are $\frac{1}{3}$ and $\frac{1}{5}$ closer to 0 or 1 whole? What is your estimation of $\frac{1}{3} + \frac{1}{5}$? More than 1 or less than 1?
- Is $\frac{7}{8}$ closer to 0 or 1 whole?
- What's $6 + \frac{7}{8}$ equal to?

NOTES ON MULTIPLE MEANS OF ACTION AND EXPRESSION:

English language learners and students with disabilities might require more examples and more time to process. If necessary, when the class is working on classwork independently, pull out a small group to do more examples.

Allow students to use the actual fraction pieces to estimate, if available. If not, allow them to draw the fractions on personal white boards.

Problem 8: $4\frac{9}{10} - 1\frac{1}{8}$ ____ $2\frac{1}{2} + \frac{2}{7}$

Have students work with partners or individually for the last problem, and then review as a class.

Problem Set (10 minutes)

Students should do their personal best to complete the Problem Set within the allotted 10 minutes. For some classes, it may be appropriate to modify the assignment by specifying which problems they work on first. Some problems do not specify a method for solving. Students should solve these problems using the RDW approach used for Application Problems.

Student Debrief (10 minutes)

Lesson Objective: Use fraction benchmark numbers to assess reasonableness of addition and subtraction equations.

The Student Debrief is intended to invite reflection and active processing of the total lesson experience.

Invite students to review their solutions for the Problem Set. They should check work by comparing answers with a partner before going over answers as a class. Look for misconceptions or misunderstandings that can be addressed in the Debrief. Guide students in a conversation to debrief the Problem Set and process the lesson.

T: Bring your Problem Set to the Debrief. Share, check, and/or explain your answers to your partner.

S: (Work together for 2 minutes.)

T: (Circulate and listen to explanations. Analyze the work to determine which student solutions to display to support the lesson objective.)

T: (Review answers or select individual students to explain the thinking process that led them to a correct answer.)

T: What did you learn today? Turn and share with your partner.

S: I can use my reasoning skills to estimate fraction answers. → When I estimate fraction answers, I should be thinking about if that fraction is closer to 0, $\frac{1}{2}$, or 1 whole. That'll make it easier for me to do mental math. → I learned to estimate fractions and answers mentally. It reminds me of rounding. → If I'm adding 2 fractions that are more than $\frac{1}{2}$, then the answer will be more than 1 whole.

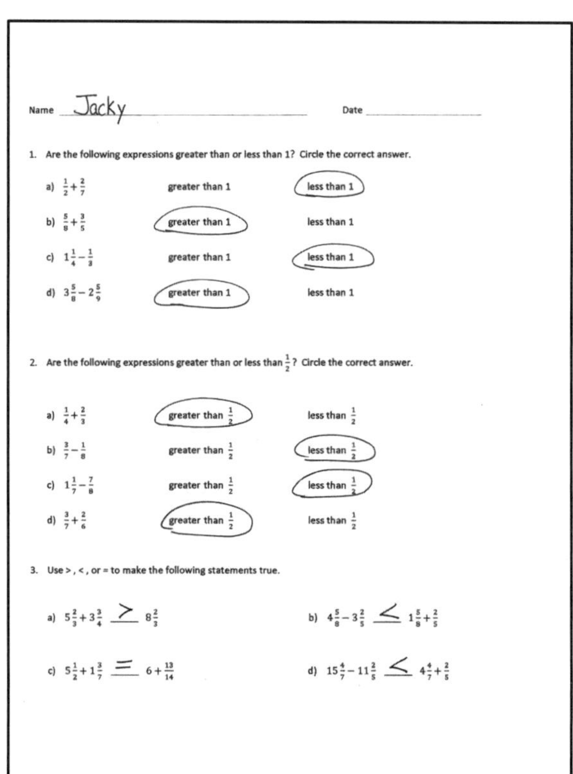
Name Jacky Date

1. Are the following expressions greater than or less than 1? Circle the correct answer.

a) $\frac{1}{2}+\frac{2}{7}$ greater than 1 (less than 1)

b) $\frac{5}{8}+\frac{3}{5}$ (greater than 1) less than 1

c) $1\frac{1}{4}-\frac{1}{3}$ greater than 1 (less than 1)

d) $3\frac{5}{8}-2\frac{5}{9}$ (greater than 1) less than 1

2. Are the following expressions greater than or less than $\frac{1}{2}$? Circle the correct answer.

a) $\frac{1}{4}+\frac{2}{3}$ (greater than $\frac{1}{2}$) less than $\frac{1}{2}$

b) $\frac{3}{7}-\frac{1}{8}$ greater than $\frac{1}{2}$ (less than $\frac{1}{2}$)

c) $1\frac{1}{7}-\frac{7}{8}$ greater than $\frac{1}{2}$ (less than $\frac{1}{2}$)

d) $\frac{3}{7}+\frac{2}{6}$ (greater than $\frac{1}{2}$) less than $\frac{1}{2}$

3. Use >, <, or = to make the following statements true.

a) $5\frac{2}{3}+3\frac{3}{4}$ __>__ $8\frac{2}{3}$

b) $4\frac{5}{8}-3\frac{2}{5}$ __<__ $1\frac{5}{8}+\frac{2}{5}$

c) $5\frac{1}{2}+1\frac{3}{7}$ __=__ $6+\frac{13}{14}$

d) $15\frac{4}{7}-11\frac{2}{5}$ __<__ $4\frac{4}{7}+\frac{2}{5}$

Note: (Optional as time allows.) The following is a suggested list of questions to invite reflection and active processing of the total lesson experience. Use those that best support students' ability to articulate the focus of the lesson.)

- Why do mathematicians agree it is wise to estimate *before* calculating?
- Think about what happens to your reasoning when you are calculating.
- Why do mathematicians agree it is wise to estimate *after* calculating?

4. Is it true that $4\frac{3}{5} - 3\frac{2}{3} = 1 + \frac{3}{5} + \frac{2}{3}$? Prove your answer.

$4\frac{3}{5} - 3\frac{2}{3}$
$= 1\frac{3}{5} - \frac{2}{3}$
$= 1 + \frac{3}{5} - \frac{2}{3}$

$1 + \frac{3}{5} + \frac{2}{3} \neq 1 + \frac{3}{5} - \frac{2}{3}$

No! It's not true! It's $\frac{2}{3}$ less, not more.

5. Jackson needs to be $1\frac{3}{4}$ inches taller in order to ride the roller coaster. Since he can't wait, he puts on a pair of boots that add $1\frac{1}{6}$ inches to his height, and slips an insole inside to add another $\frac{1}{8}$ inches to his height. Will this make Jackson appear tall enough to ride the roller coaster?

Is $1\frac{1}{6} + \frac{1}{8}$ greater than or equal to $1\frac{3}{4}$?

Since both $\frac{1}{6}$ and $\frac{1}{8}$ is less than $\frac{1}{2}$ or $\frac{2}{4}$, then it is less than $\frac{3}{4}$. So, $1\frac{1}{6} + \frac{1}{8} < 1\frac{3}{4}$. No. The boots and the insole will not make Jackson appear tall enough to ride the roller coaster.

6. A baker needs 5 lb of butter for a recipe. She found 2 portions that each weigh $1\frac{1}{6}$ lb and a portion that weighs $2\frac{2}{7}$ lb. Does she have enough butter for her recipe?

Is $1\frac{1}{6} + 1\frac{1}{6} + 2\frac{2}{7} > 5$?

$1 + 1 + 2 + \frac{2}{6} + \frac{2}{7}$

$4 + \underbrace{\frac{2}{6} + \frac{2}{7}}_{\text{less than } 1}$

So, $4 + \frac{2}{6} + \frac{2}{7} < 5$.

No. The baker does not have enough butter for her recipe.

Exit Ticket (3 minutes)

After the Student Debrief, instruct students to complete the Exit Ticket. A review of their work will help with assessing students' understanding of the concepts that were presented in today's lesson and planning more effectively for future lessons. The questions may be read aloud to the students.

Name ______________________________ Date ______________

1. Are the following expressions greater than or less than 1? Circle the correct answer.

a. $\frac{1}{2}+\frac{2}{7}$ greater than 1 less than 1

b. $\frac{5}{8}+\frac{3}{5}$ greater than 1 less than 1

c. $1\frac{1}{4}-\frac{1}{3}$ greater than 1 less than 1

d. $3\frac{5}{8}-2\frac{5}{9}$ greater than 1 less than 1

2. Are the following expressions greater than or less than $\frac{1}{2}$? Circle the correct answer.

a. $\frac{1}{4}+\frac{2}{3}$ greater than $\frac{1}{2}$ less than $\frac{1}{2}$

b. $\frac{3}{7}-\frac{1}{8}$ greater than $\frac{1}{2}$ less than $\frac{1}{2}$

c. $1\frac{1}{7}-\frac{7}{8}$ greater than $\frac{1}{2}$ less than $\frac{1}{2}$

d. $\frac{3}{7}+\frac{2}{6}$ greater than $\frac{1}{2}$ less than $\frac{1}{2}$

3. Use >, <, or = to make the following statements true.

a. $5\frac{2}{3}+3\frac{3}{4}$ ______ $8\frac{2}{3}$

b. $4\frac{5}{8}-3\frac{2}{5}$ ______ $1\frac{5}{8}+\frac{2}{5}$

c. $5\frac{1}{2}+1\frac{3}{7}$ ______ $6+\frac{13}{14}$

d. $15\frac{4}{7}-11\frac{2}{5}$ ______ $4\frac{4}{7}+\frac{2}{5}$

4. Is it true that $4\frac{3}{5} - 3\frac{2}{3} = 1 + \frac{3}{5} + \frac{2}{3}$? Prove your answer.

5. Jackson needs to be $1\frac{3}{4}$ inches taller in order to ride the roller coaster. Since he can't wait, he puts on a pair of boots that add $1\frac{1}{6}$ inches to his height and slips an insole inside to add another $\frac{1}{8}$ inch to his height. Will this make Jackson appear tall enough to ride the roller coaster?

6. A baker needs 5 lb of butter for a recipe. She found 2 portions that each weigh $1\frac{1}{6}$ lb and a portion that weighs $2\frac{2}{7}$ lb. Does she have enough butter for her recipe?

Name ______________________________ Date ______________

1. Circle the correct answer.

a. $\frac{1}{2} + \frac{5}{12}$ greater than 1 less than 1

b. $2\frac{7}{8} - 1\frac{7}{9}$ greater than 1 less than 1

c. $1\frac{1}{12} - \frac{7}{10}$ greater than $\frac{1}{2}$ less than $\frac{1}{2}$

d. $\frac{3}{7} + \frac{1}{8}$ greater than $\frac{1}{2}$ less than $\frac{1}{2}$

2. Use > , < , or = to make the following statement true.

$4\frac{4}{5} + 3\frac{2}{3}$ ______ $8\frac{1}{2}$

Name ______________________________ Date ______________

1. Are the following expressions greater than or less than 1? Circle the correct answer.

a. $\frac{1}{2}+\frac{4}{9}$ greater than 1 less than 1

b. $\frac{5}{8}+\frac{3}{5}$ greater than 1 less than 1

c. $1\frac{1}{5}-\frac{1}{3}$ greater than 1 less than 1

d. $4\frac{3}{5}-3\frac{3}{4}$ greater than 1 less than 1

2. Are the following expressions greater than or less than $\frac{1}{2}$? Circle the correct answer.

a. $\frac{1}{5}+\frac{1}{4}$ greater than $\frac{1}{2}$ less than $\frac{1}{2}$

b. $\frac{6}{7}-\frac{1}{6}$ greater than $\frac{1}{2}$ less than $\frac{1}{2}$

c. $1\frac{1}{7}-\frac{5}{6}$ greater than $\frac{1}{2}$ less than $\frac{1}{2}$

d. $\frac{4}{7}+\frac{1}{8}$ greater than $\frac{1}{2}$ less than $\frac{1}{2}$

3. Use >, <, or = to make the following statements true.

a. $5\frac{4}{5}+2\frac{2}{3}$ ______ $8\frac{3}{4}$

b. $3\frac{4}{7}-2\frac{3}{5}$ ______ $1\frac{4}{7}+\frac{3}{5}$

c. $4\frac{1}{2}+1\frac{4}{9}$ ______ $5+\frac{13}{18}$

d. $10\frac{3}{8}-7\frac{3}{5}$ ______ $3\frac{3}{8}+\frac{3}{5}$

4. Is it true that $5\frac{2}{3} - 3\frac{3}{4} = 1 + \frac{2}{3} + \frac{3}{4}$? Prove your answer.

5. A tree limb hangs $5\frac{1}{4}$ feet from a telephone wire. The city trims back the branch *before* it grows within $2\frac{1}{2}$ feet of the wire. Will the city allow the tree to grow $2\frac{3}{4}$ more feet?

6. Mr. Kreider wants to paint two doors and several shutters. It takes $2\frac{1}{8}$ gallons of paint to coat each door and $1\frac{3}{5}$ gallons of paint to coat all of his shutters. If Mr. Kreider buys three 2-gallon cans of paint, does he have enough to complete the job?

Lesson 14

Objective: Strategize to solve multi-term problems.

Suggested Lesson Structure

Fluency Practice	(13 minutes)
Application Problems	(10 minutes)
Concept Development	(27 minutes)
Student Debrief	(10 minutes)
Total Time	**(60 minutes)**

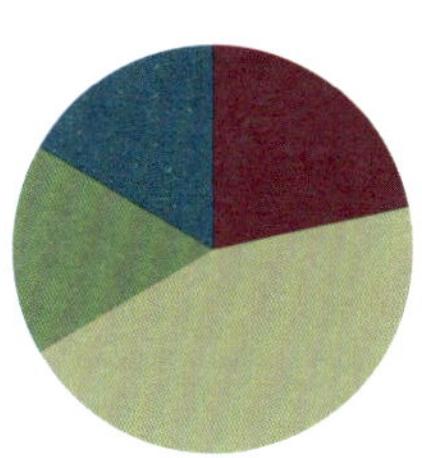

Fluency Practice (13 minutes)

- Sprint: Make Larger Units **4.NF.1** (10 minutes)
- Happy Counting with Mixed Numbers **4.NF.3a** (3 minutes)

Sprint: Make Larger Units (10 minutes)

Materials: (S) Make Larger Units Sprint

Note: This Sprint reviews making like units, a prerequisite skill for today's work with multi-term problems.

NOTES ON MULTIPLE MEANS OF ENGAGEMENT:

Success with this Sprint depends greatly on students' knowledge of their factors. Students working below grade level often are not fluent with their basic facts. The day before administering this Sprint, discreetly meet with students to give them a copy of the next day's Sprint. This is very motivating. Now, they have a reason to study and practice.

Happy Counting with Mixed Numbers (3 minutes)

Note: This activity builds comfort and fluency with mixed numbers.

T: Let's count by $\frac{1}{2}$ with mixed numbers. Ready? (Rhythmically point up until a change is desired. Show a closed hand, and then point down. Continue, mixing it up).

S: $\frac{1}{2}$, 1, $1\frac{1}{2}$, 2 (stop), $1\frac{1}{2}$, 1, $\frac{1}{2}$, 0 (stop), $\frac{1}{2}$, 1, $1\frac{1}{2}$, 2, $2\frac{1}{2}$, 3, $3\frac{1}{2}$, 4 (stop), $3\frac{1}{2}$, 3, $2\frac{1}{2}$, 2, $1\frac{1}{2}$, 1 (stop), $1\frac{1}{2}$, 2, $2\frac{1}{2}$, 3, $3\frac{1}{2}$, 4, $4\frac{1}{2}$, 5.

T: Excellent. Try it for 30 seconds with your partner. Partner A, you are the teacher today.

Application Problems (10 minutes)

For a large order, Mr. Magoo made $\frac{3}{8}$ kg of fudge in his bakery. He then got $\frac{1}{6}$ kg from his sister's bakery. If he needs a total of $1\frac{1}{2}$ kg, how much more fudge does he need to make?

During lunch, Charlie drinks $2\frac{3}{4}$ cups of milk. Allison drinks $\frac{3}{8}$ cup of milk. Carmen drinks $1\frac{1}{6}$ cups of milk. How much milk do the 3 students drink?

$1\frac{1}{2}$ kg

$\frac{3}{8}$ kg	$\frac{1}{6}$ kg	needs
B	S	?

$\left(1\frac{1}{2}\text{ kg} - \frac{3}{8}\text{ kg}\right) - \frac{1}{6}\text{ kg} = \_\_\_$

$1\frac{4}{8} - \frac{3}{8}$

$= 1\frac{1}{8}$

$1\frac{1}{8} - \frac{1}{6}$

$= 1\frac{3}{24} - \frac{4}{24}$

$= \frac{23}{24}$

Mr. Magoo needs $\frac{23}{24}$ kg more fudge.

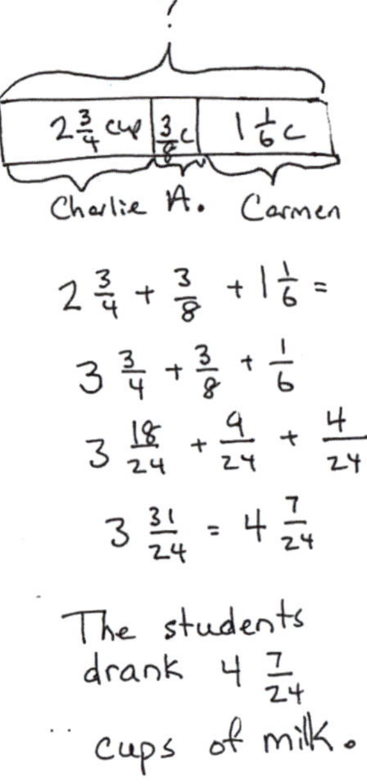

T: Now that you have solved these two problems, consider how they are the same and how they are different.

S: Both problems had three parts that we knew. → True, but actually, in the fudge problem, the one part was the whole amount. → The fudge problem had a missing part, but the milk problem was missing the whole amount of milk. → So, for the fudge problem, we had to subtract from $1\frac{1}{2}$ kg. For the milk problem, we had to add up the three parts to find the total amount of milk.

Note: These multi-step Application Problems ask students to make like units to add and subtract as an anticipatory set for today's Concept Development.

Concept Development (27 minutes)

Materials: (S) Personal white board

Problem 1: $\frac{2}{3} + \frac{1}{5} + \frac{1}{3} + 1\frac{4}{5}$

T: Look at this problem. What do you notice? Turn and share with a partner.

S: I see that it's an addition problem adding thirds and fifths. → I see that I can add up the thirds, and I can also add the fifths together.

T: Can you solve this problem mentally? Turn and share.

S: (Share.)

T: Are we finding a part or whole? How will you solve?

S: Whole. → $\frac{2}{3} + \frac{1}{3}$ = 1 whole. $\frac{1}{5} + 1\frac{4}{5}$ = 2 wholes. Finally, 1 + 2 = 3.

T: Excellent. We can rearrange the problem and solve it using that strategy.

$$\frac{2}{3} + \frac{1}{5} + \frac{1}{3} + 1\frac{4}{5}$$
$$= \left(\frac{2}{3} + \frac{1}{3}\right) + \left(\frac{1}{5} + 1\frac{4}{5}\right)$$
$$= 1 + 2$$
$$= 3$$

Problem 2: $5\frac{7}{8} - \frac{1}{2} - \frac{7}{8} - 1\frac{1}{2}$

T: Analyze this problem with your neighbor.

S: I see that it's a subtraction problem. → I see that denominators are eighths and halves. They need to be the same or like units for me to subtract. → Without looking at the mixed numbers, I see two 7 eighths and two 1 halves.

$$5\frac{7}{8} - \frac{1}{2} - \frac{7}{8} - 1\frac{1}{2}$$
$$= \left(5\frac{7}{8} - \frac{7}{8}\right) - \left(\frac{1}{2} + 1\frac{1}{2}\right)$$
$$= 5 - 2$$
$$= 3$$

$$5\frac{7}{8} - \frac{1}{2} - \frac{7}{8} - 1\frac{1}{2}$$
$$= 5\frac{7}{8} - \left(\frac{1}{2} + \frac{7}{8} + 1\frac{1}{2}\right)$$
$$= 5\frac{7}{8} - \left(\frac{4}{8} + \frac{7}{8} + 1\frac{4}{8}\right)$$
$$= 5\frac{7}{8} - 2\frac{7}{8}$$
$$= 3$$

T: Yes. This is a subtraction problem. Analyze the parts and wholes. Turn and share.

S: $5\frac{7}{8}$ is the whole amount. $\frac{7}{8}$ is a part being taken away. That makes 5. → $1\frac{1}{2}$ and $\frac{1}{2}$ are both parts being taken away. If I combine them, I'm taking away 2. 5 – 2 = 3. → We can combine all the parts and make a larger part, and then subtract from the whole.

NOTES ON MULTIPLE MEANS OF ENGAGEMENT:

When students are analyzing parts and wholes, relationships, or compatible numbers, resist the temptation to jump in. Wait time is critical. Let them analyze. This allows students working above grade level to find more complexities and those students working below grade level to have more think time.

Problem 3 is more challenging because of the change in sign. Students see the like units of $2\frac{5}{6}$ and $\frac{1}{6}$, which expedites the addition of the two numbers to make a larger whole from which one third is subtracted.

Problem 3: $2\frac{5}{6} - \frac{1}{3} + \frac{1}{6}$

Follow a similar sequence of analysis and solution as shown below.

$$2\frac{5}{6} - \frac{1}{3} + \frac{1}{6}$$
$$= \left(2\frac{5}{6} + \frac{1}{6}\right) - \frac{1}{3}$$
$$= 3 - \frac{1}{3}$$
$$= 2\frac{2}{3}$$

Problem 4: $\frac{14}{3} +$ ____ $+ \frac{9}{4} = 8\frac{11}{12}$

T: Let's analyze this fraction equation. Tell your partner what you notice.

S: This is an addition problem, and the sum of $8\frac{11}{12}$ is on the right-hand side. → I'm missing a part that is needed to make the total amount of $8\frac{11}{12}$. → $\frac{14}{3}$ is a part, too. → I can add the parts and subtract them from the whole amount to find that mystery number. → Find the sum of the parts and take them away from the whole.

T: Go ahead and solve for the missing part. You can use paper and pencil if you wish.

$$\frac{14}{3} + \_\_\_\_ + \frac{9}{4} = 8\frac{11}{12}$$
$$= 8\frac{11}{12} - \frac{14}{3} - \frac{9}{4}$$
$$= 8\frac{11}{12} - \left(\frac{14}{3} + \frac{9}{4}\right)$$
$$= 8\frac{11}{12} - \left(4\frac{2}{3} + 2\frac{1}{4}\right)$$
$$= 8\frac{11}{12} - 6\frac{11}{12}$$
$$= 2$$

Problem 5: ____ $- 15 - 4\frac{1}{2} = 7\frac{3}{5}$

T: Tell your neighbor what you know about this problem.

S: I see that it's a subtraction problem. Something minus 15 minus $4\frac{1}{2}$ equals $7\frac{3}{5}$. → The whole is missing in this problem, and everything else is a part. → I can add up all the parts together to find the whole. → The whole is missing, so we'll add up all the parts to find the whole.

$$\_\_\_\_ - 15 - 4\frac{1}{2} = 7\frac{3}{5}$$
$$15 + 4\frac{1}{2} + 7\frac{3}{5} = \_\_\_\_$$
$$= 15 + 4\frac{5}{10} + 7\frac{6}{10}$$
$$= 26\frac{11}{10}$$
$$= 27\frac{1}{10}$$

EUREKA MATH™

Problem 6: $6\frac{3}{4} + \frac{3}{5} - \_\_\_\_ = 5$

T: I would like you to try to solve this problem with your partner. You have two minutes to talk about the problem and determine your strategy.

Allow two minutes for students to analyze and discuss the problem without calculating, just formulating their thoughts about how to solve.

T: Go ahead and solve the problem.

S: (Solve.)

$$6\frac{3}{4} + \frac{3}{5} - \_\_\_\_ = 5$$
$$= \left(6\frac{3}{4} + \frac{3}{5}\right) - 5$$
$$= 6\frac{15}{20} + \frac{12}{20} - 5$$
$$= 6\frac{27}{20} - 5$$
$$= 7\frac{7}{20} - 5$$
$$= 2\frac{7}{20}$$

NOTES ON MULTIPLE MEANS OF ACTION AND EXPRESSION:

The problems on this particular Problem Set may require more room than the Problem Set offers. Be aware that students do well to have a math notebook or journal. When a Problem Set has a set of challenging problems, assign pairs to solve them on the board as others use paper so that they are easier to review. It is also much more engaging for students to see their peers' solutions.

Problem Set (10 minutes)

Students should do their personal best to complete the Problem Set within the allotted 10 minutes. For some classes, it may be appropriate to modify the assignment by specifying which problems they work on first. Some problems do not specify a method for solving. Students should solve these problems using the RDW approach used for Application Problems.

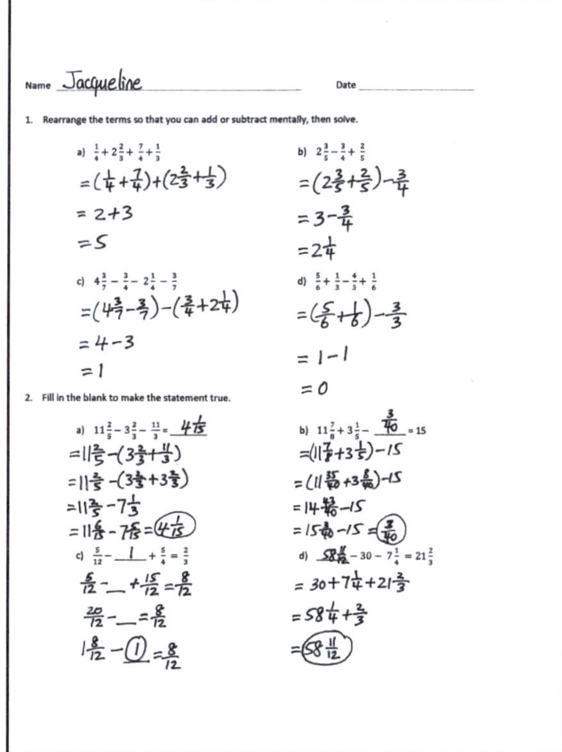

Name Jacqueline Date

1. Rearrange the terms so that you can add or subtract mentally, then solve.

a) $\frac{1}{4} + 2\frac{2}{3} + \frac{7}{4} + \frac{1}{3}$

$= \left(\frac{1}{4} + \frac{7}{4}\right) + \left(2\frac{2}{3} + \frac{1}{3}\right)$

$= 2 + 3$

$= 5$

b) $2\frac{3}{5} - \frac{3}{4} + \frac{2}{5}$

$= \left(2\frac{3}{5} + \frac{2}{5}\right) - \frac{3}{4}$

$= 3 - \frac{3}{4}$

$= 2\frac{1}{4}$

c) $4\frac{3}{7} - \frac{3}{4} - 2\frac{1}{4} - \frac{3}{7}$

$= \left(4\frac{3}{7} - \frac{3}{7}\right) - \left(\frac{3}{4} + 2\frac{1}{4}\right)$

$= 4 - 3$

$= 1$

d) $\frac{5}{6} + \frac{1}{3} - \frac{4}{3} + \frac{1}{6}$

$= \left(\frac{5}{6} + \frac{1}{6}\right) - \frac{3}{3}$

$= 1 - 1$

$= 0$

2. Fill in the blank to make the statement true.

a) $11\frac{2}{5} - 3\frac{2}{3} - \frac{11}{3} =$ $4\frac{1}{15}$

$= 11\frac{2}{5} - \left(3\frac{2}{3} + \frac{11}{3}\right)$

$= 11\frac{2}{5} - \left(3\frac{2}{3} + 3\frac{2}{3}\right)$

$= 11\frac{2}{5} - 7\frac{1}{3}$

$= 11\frac{6}{15} - 7\frac{5}{15} = 4\frac{1}{15}$

b) $11\frac{7}{8} + 3\frac{1}{5} -$ $\frac{3}{40}$ $= 15$

$= \left(11\frac{7}{8} + 3\frac{1}{5}\right) - 15$

$= \left(11\frac{35}{40} + 3\frac{8}{40}\right) - 15$

$= 14\frac{43}{40} - 15$

$= 15\frac{3}{40} - 15 = \frac{3}{40}$

c) $\frac{5}{12} -$ 1 $+ \frac{5}{4} = \frac{2}{3}$

$\frac{5}{12} - \_\_ + \frac{15}{12} = \frac{8}{12}$

$\frac{20}{12} - \_\_ = \frac{8}{12}$

$1\frac{8}{12} - 1 = \frac{8}{12}$

d) $58\frac{11}{12}$ $- 30 - 7\frac{1}{4} = 21\frac{2}{3}$

$= 30 + 7\frac{1}{4} + 21\frac{2}{3}$

$= 58\frac{1}{4} + \frac{2}{3}$

$= 58\frac{11}{12}$

Student Debrief (10 minutes)

Lesson Objective: Strategize to solve multi-term problems.

The Student Debrief is intended to invite reflection and active processing of the total lesson experience.

Invite students to review their solutions for the Problem Set. They should check work by comparing answers with a partner before going over answers as a class. Look for misconceptions or misunderstandings that can be addressed in the Debrief. Guide students in a conversation to debrief the Problem Set and process the lesson.

T: When rearranging the terms in the top section of the Problem Set, talk to your partner about what you looked for to help you solve the problems easily.

S: I grouped to make whole numbers. → By grouping fractions to make whole numbers, it was easy to add or subtract. → I looked for numbers in different forms. It was harder to see the pairs if the denominators were different or if they weren't written as mixed numbers. → Problem 1(c) was so much harder! I was really surprised the answer was 1. I didn't expect that, so it made me go back and look at the relationships in the problem.

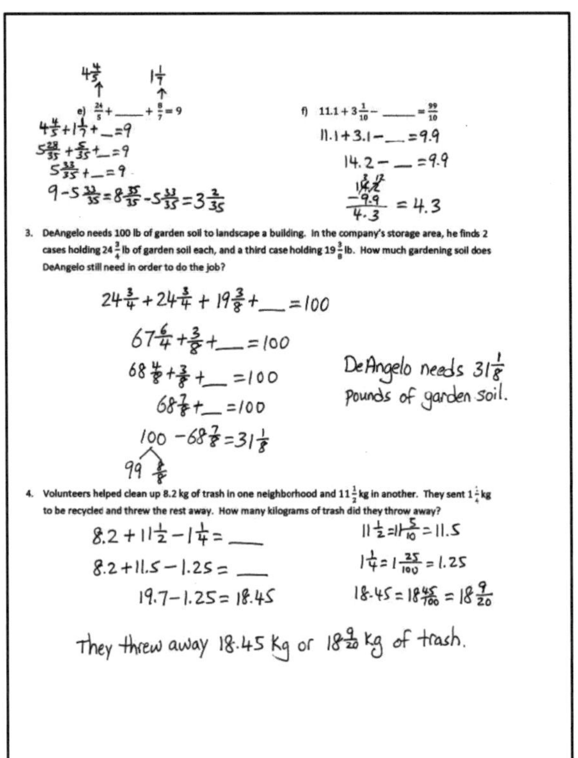

T: Talk to your partner about some of the skills you had to use to solve these problems.

S: We had to analyze part and whole relationships. → I had to recognize when there were easy like units. → We had to move back and forth between decimals and fractions in Problem 2(f) and in the second word problem about the volunteers, too. → We had to think hard about the problems that had addition and subtraction problems and whether to add or subtract something.

T: This was a challenging activity.

S: We had to really think!

T: Let's go over the last problem about the volunteers. I would say it is related to Problem 1(b). Explain my thinking to your partner.

T: (After students talk.) Which of the earlier problems on the Problem Set would you relate to the problem of the gardening soil?

T: (After students talk.) Review the strategies you used on two problems. First, review a problem that was very easy for you. Then, review the strategy on a problem that was very challenging for you.

Exit Ticket (3 minutes)

After the Student Debrief, instruct students to complete the Exit Ticket. A review of their work will help with assessing students' understanding of the concepts that were presented in today's lesson and planning more effectively for future lessons. The questions may be read aloud to the students.

A

Number Correct: _______

Make Larger Units

1.	$^{2}/_{4}$ =			23.	$^{9}/_{27}$ =	
2.	$^{2}/_{6}$ =			24.	$^{9}/_{63}$ =	
3.	$^{2}/_{8}$ =			25.	$^{8}/_{12}$ =	
4.	$^{5}/_{10}$ =			26.	$^{8}/_{16}$ =	
5.	$^{5}/_{15}$ =			27.	$^{8}/_{24}$ =	
6.	$^{5}/_{20}$ =			28.	$^{8}/_{64}$ =	
7.	$^{4}/_{8}$ =			29.	$^{12}/_{18}$ =	
8.	$^{4}/_{12}$ =			30.	$^{12}/_{16}$ =	
9.	$^{4}/_{16}$ =			31.	$^{9}/_{12}$ =	
10.	$^{3}/_{6}$ =			32.	$^{6}/_{8}$ =	
11.	$^{3}/_{9}$ =			33.	$^{10}/_{12}$ =	
12.	$^{3}/_{12}$ =			34.	$^{15}/_{18}$ =	
13.	$^{4}/_{6}$ =			35.	$^{8}/_{10}$ =	
14.	$^{6}/_{12}$ =			36.	$^{16}/_{20}$ =	
15.	$^{6}/_{18}$ =			37.	$^{12}/_{15}$ =	
16.	$^{6}/_{30}$ =			38.	$^{18}/_{27}$ =	
17.	$^{6}/_{9}$ =			39.	$^{27}/_{36}$ =	
18.	$^{7}/_{14}$ =			40.	$^{32}/_{40}$ =	
19.	$^{7}/_{21}$ =			41.	$^{45}/_{54}$ =	
20.	$^{7}/_{42}$ =			42.	$^{24}/_{36}$ =	
21.	$^{8}/_{12}$ =			43.	$^{60}/_{72}$ =	
22.	$^{9}/_{18}$ =			44.	$^{48}/_{60}$ =	

B

Number Correct: _______

Improvement: _______

Make Larger Units

1.	$^{5}/_{10} =$		23.	$^{8}/_{24} =$	
2.	$^{5}/_{15} =$		24.	$^{8}/_{56} =$	
3.	$^{5}/_{20} =$		25.	$^{8}/_{12} =$	
4.	$^{2}/_{4} =$		26.	$^{9}/_{18} =$	
5.	$^{2}/_{6} =$		27.	$^{9}/_{27} =$	
6.	$^{2}/_{8} =$		28.	$^{9}/_{72} =$	
7.	$^{3}/_{6} =$		29.	$^{12}/_{18} =$	
8.	$^{3}/_{9} =$		30.	$^{6}/_{8} =$	
9.	$^{3}/_{12} =$		31.	$^{9}/_{12} =$	
10.	$^{4}/_{8} =$		32.	$^{12}/_{16} =$	
11.	$^{4}/_{12} =$		33.	$^{8}/_{10} =$	
12.	$^{4}/_{16} =$		34.	$^{16}/_{20} =$	
13.	$^{4}/_{6} =$		35.	$^{12}/_{15} =$	
14.	$^{7}/_{14} =$		36.	$^{10}/_{12} =$	
15.	$^{7}/_{21} =$		37.	$^{15}/_{18} =$	
16.	$^{7}/_{35} =$		38.	$^{16}/_{24} =$	
17.	$^{6}/_{9} =$		39.	$^{24}/_{32} =$	
18.	$^{6}/_{12} =$		40.	$^{36}/_{45} =$	
19.	$^{6}/_{18} =$		41.	$^{40}/_{48} =$	
20.	$^{6}/_{36} =$		42.	$^{24}/_{36} =$	
21.	$^{8}/_{12} =$		43.	$^{48}/_{60} =$	
22.	$^{8}/_{16} =$		44.	$^{60}/_{72} =$	

Name ______________________________ Date ________________

1. Rearrange the terms so that you can add or subtract mentally. Then, solve.

   a. $\frac{1}{4} + 2\frac{2}{3} + \frac{7}{4} + \frac{1}{3}$

   b. $2\frac{3}{5} - \frac{3}{4} + \frac{2}{5}$

   c. $4\frac{3}{7} - \frac{3}{4} - 2\frac{1}{4} - \frac{3}{7}$

   d. $\frac{5}{6} + \frac{1}{3} - \frac{4}{3} + \frac{1}{6}$

2. Fill in the blank to make the statement true.

   a. $11\frac{2}{5} - 3\frac{2}{3} - \frac{11}{3} =$ ______

   b. $11\frac{7}{8} + 3\frac{1}{5} -$ ______ $= 15$

c. $\frac{5}{12} - \_\_\_\_\_\_ + \frac{5}{4} = \frac{2}{3}$

d. $\_\_\_\_\_\_ - 30 - 7\frac{1}{4} = 21\frac{2}{3}$

e. $\frac{24}{5} + \_\_\_\_\_\_ + \frac{8}{7} = 9$

f. $11.1 + 3\frac{1}{10} - \_\_\_\_\_\_ = \frac{99}{10}$

3. DeAngelo needs 100 lb of garden soil to landscape a building. In the company's storage area, he finds 2 cases holding $24\frac{3}{4}$ lb of garden soil each, and a third case holding $19\frac{3}{8}$ lb. How much gardening soil does DeAngelo still need in order to do the job?

4. Volunteers helped clean up 8.2 kg of trash in one neighborhood and $11\frac{1}{2}$ kg in another. They sent $1\frac{1}{4}$ kg to be recycled and threw the rest away. How many kilograms of trash did they throw away?

Name ______________________________ Date ______________

Fill in the blank to make the statement true.

1. $1\frac{3}{4} + \frac{1}{6} + \_\_\_\_\_ = 7\frac{1}{2}$

2. $8\frac{4}{5} - \frac{2}{3} - \_\_\_\_\_ = 3\frac{1}{10}$

Name ______________________________ Date ______________

1. Rearrange the terms so that you can add or subtract mentally. Then, solve.

a. $1\frac{3}{4} + \frac{1}{2} + \frac{1}{4} + \frac{1}{2}$

b. $3\frac{1}{6} - \frac{3}{4} + \frac{5}{6}$

c. $5\frac{5}{8} - 2\frac{6}{7} - \frac{2}{7} - \frac{5}{8}$

d. $\frac{7}{9} + \frac{1}{2} - \frac{3}{2} + \frac{2}{9}$

2. Fill in the blank to make the statement true.

a. $7\frac{3}{4} - 1\frac{2}{7} - \frac{3}{2} =$ ______

b. $9\frac{5}{6} + 1\frac{1}{4} +$ ______ $= 14$

c. $\frac{7}{10} - \_\_\_\_\_\_ + \frac{3}{2} = \frac{6}{5}$

d. $\_\_\_\_\_\_ - 20 - 3\frac{1}{4} = 14\frac{5}{8}$

e. $\frac{17}{3} + \_\_\_\_\_\_ + \frac{5}{2} = 10\frac{4}{5}$

f. $23.1 + 1\frac{7}{10} - \_\_\_\_\_\_ = \frac{66}{10}$

3. Laura bought $8\frac{3}{10}$ yd of ribbon. She used $1\frac{2}{5}$ yd to tie a package and $2\frac{1}{3}$ yd to make a bow. Joe later gave her $4\frac{3}{5}$ yd. How much ribbon does she now have?

4. Mia bought $10\frac{1}{9}$ lb of flour. She used $2\frac{3}{4}$ lb of flour to bake banana cakes and some to bake chocolate cakes. After baking all the cakes, she had $3\frac{5}{6}$ lb of flour left. How much flour did she use to bake the chocolate cakes?

Lesson 15

Objective: Solve multi-step word problems; assess reasonableness of solutions using benchmark numbers.

Suggested Lesson Structure

■ Fluency Practice	(12 minutes)
■ Concept Development	(38 minutes)
■ Student Debrief	(10 minutes)
Total Time	**(60 minutes)**

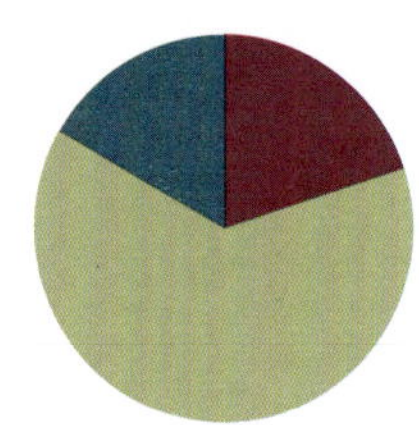

Fluency Practice (12 minutes)

- Sprint: Circle the Smaller Fraction **4.NF.2** (12 minutes)

Sprint: Circle the Smaller Fraction (12 minutes)

Materials: (S) Circle the Smaller Fraction Sprint

Note: Students practice analyzing fractions in preparation for today's task of assessing the reasonableness of a solution.

Concept Development (38 minutes)

Materials: (S) Problem Set, personal white board

Note: For this lesson, the Problem Set comprises word problems from the Concept Development and is therefore to be used during the lesson itself.

Problem 1

In a race, the second-place finisher crossed the finish line $1\frac{1}{3}$ minutes after the winner. The third-place finisher was $1\frac{3}{4}$ minutes behind the second-place finisher. The third-place finisher took $34\frac{2}{3}$ minutes. How long did the winner take?

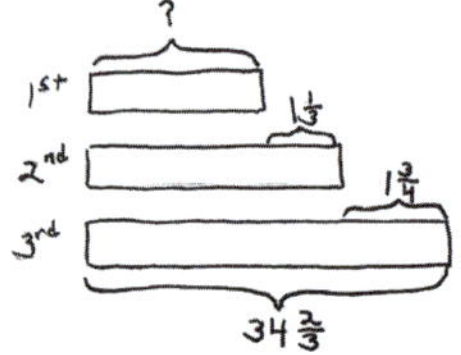

The 1st place time was 31 min 35s.

$34\frac{2}{3} - 1\frac{3}{4} = 33\frac{2}{3} - \frac{3}{4}$
$= 33\frac{8}{12} - \frac{9}{12}$
$= 32\frac{20}{12} - \frac{9}{12}$
$= 32\frac{11}{12}$

$32\frac{11}{12} - 1\frac{1}{3} = 31\frac{11}{12} - \frac{1}{3}$
$= 31\frac{11}{12} - \frac{4}{12}$
$= 31\frac{7}{12}$

$31\frac{7}{12}$ min $= 31\frac{35}{60}$ min $= 31$ min 35 s

T: Let's read the problem together.

S: (Read chorally.)

T: Now, share with your partner: What do you see when you hear the story? Explain how you are going to draw this problem.

S: (Share and explain.)

T: Ming, could you share your method of drawing?

S: The first sentence tells me that the second finisher took $1\frac{1}{3}$ minutes longer than the winner. So, I'll draw 2 bars. The second bar represents the second finisher with a longer bar and with the difference of $1\frac{1}{3}$ minutes.

T: Betty, can you add more to Ming's drawing?

S: The second sentence says the third finisher took $1\frac{3}{4}$ minutes longer than the second finisher. So, I'll draw a longer bar for the third finisher, and label the difference of $1\frac{3}{4}$ minutes.

T: Steven, can you add anything else to the drawing?

S: The third sentence tells us the third finisher's time in minutes. So, I can label the third bar with $34\frac{2}{3}$ minutes.

NOTES ON MULTIPLE MEANS OF ENGAGEMENT:

When doing a word problem lesson, be sure to provide many opportunities for students to turn and talk or repeat what a peer has said.

If possible, pair students that speak the same first language together. For example, a non-English language learner Chinese student with a strong math background can be paired with an English language learner Chinese student. Teachers can also encourage them to converse in Chinese.

MP.5

T: Excellent. The question now is to find the winner's time. How are you going to solve this problem? Turn and share with your partner.

S: We have to find the second finisher's time first, and then we can find the winner's time. → We know the third finisher's time but don't know the second finisher's time. We can solve it by subtracting. → Use the third finisher's time to subtract $1\frac{3}{4}$ to find the second finisher's time. Then, use the second finisher's time to subtract $1\frac{1}{3}$ to find the winner's time.

T: Great. Let's first find the second finisher's time. What's the subtraction sentence?

S: $34\frac{2}{3} - 1\frac{3}{4}$

$= 34\frac{8}{12} - 1\frac{9}{12}$

$= 33\frac{20}{12} - 1\frac{9}{12}$

$= 32\frac{11}{12}$

NOTES ON MULTIPLE MEANS OF ENGAGEMENT:

An extension activity can be used to convert the winner's time into seconds.

T: What does $32\frac{11}{12}$ mean?

S: The second finisher's time is $32\frac{11}{12}$ minutes.

T: Let's now find the winner's time. What's the subtraction sentence?

S: $32\frac{11}{12} - 1\frac{1}{3}$

$= 32\frac{11}{12} - 1\frac{4}{12}$

$= 31\frac{7}{12}$

T: What's the word sentence to answer the question?

S: The winner's time was $31\frac{7}{12}$ minutes.

T: How do I convert $31\frac{7}{12}$ minutes to minutes and seconds? Turn and share with your partner.

T: Alanzo, can you share your thinking with us?

S: $31\frac{7}{12}$ minutes means there are 31 minutes and $\frac{7}{12}$ of a minute. I need to convert $\frac{7}{12}$ into seconds.

T: Linda, what do you think?

S: I agree with Alanzo. I know there are 60 seconds in a minute, so I'll convert $\frac{7}{12}$ to $\frac{35}{60}$.

T: Very good. $\frac{7}{12} = \frac{35}{60}$. What's the winner's time in minutes and seconds?

S: The winner's time was 31 minutes and 35 seconds.

Problem 2

John used $1\frac{3}{4}$ kg of salt to melt the ice on his sidewalk. He then used another $3\frac{4}{5}$ kg on the driveway. If he originally bought 10 kg of salt, how much does he have left?

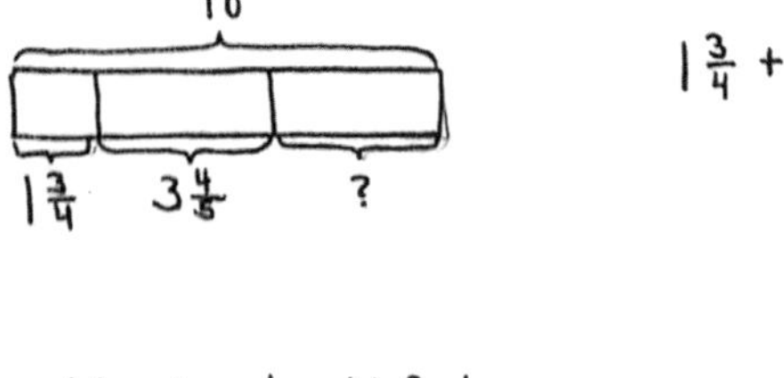

He had $4\frac{9}{20}$ kg of salt left.

$$1\frac{3}{4} + 3\frac{4}{5} = 4\frac{3}{4} + \frac{4}{5}$$
$$= 4\frac{15}{20} + \frac{16}{20}$$
$$= 4\frac{31}{20}$$
$$= 5\frac{11}{20}$$
$$10 - 5\frac{11}{20} = 5 - \frac{11}{20}$$
$$= 4\frac{9}{20}$$

T: Let's read the problem together.

S: (Read chorally.)

T: What do you see when you hear the story? How you are going to draw this problem? Turn and share.

S: (Share.)

T: I'll give you one minute to draw. Explain your conclusions to your partner based on your drawing.

S: (Discuss briefly.)

T: Student A, could you share your method of drawing?

S: Since I know he bought 10 kg of salt, I drew a whole tape and labeled it 10. He used some salt for the sidewalk and some for the driveway. I partitioned the tape diagram into 3 parts and labeled them as $1\frac{3}{4}$, $3\frac{4}{5}$, and put a question mark for the part he has left.

T: How much salt does he have left? How do we solve this problem? Turn and share.

S: I can use the total of 10 kg to subtract the two parts to find the leftover part. → I can add up the two parts to make them a larger part. Then I'll subtract that from the whole of 10 kg.

T: You have four minutes to finish solving the problem.

Problem Set (20 minutes)

Students should do their personal best to complete the Problem Set within the allotted 20 minutes. For some classes, it may be appropriate to modify the assignment by specifying which problems they work on first. Some problems do not specify a method for solving. Students should solve these problems using the RDW approach used for Application Problems.

Problem 3

Sinister Stan stole $3\frac{3}{4}$ oz of slime from Messy Molly, but his evil plans require $6\frac{3}{8}$ oz of slime. He stole another $2\frac{3}{5}$ oz from Rude Ralph. How much more slime does Sinister Stan need for his evil plan?

$6\frac{3}{8}$

$3\frac{3}{4}$ $2\frac{3}{5}$?

$3\frac{3}{4} + 2\frac{3}{5} = 5\frac{3}{4} + \frac{3}{5}$
$= 5\frac{15}{20} + \frac{12}{20}$
$= 5\frac{27}{20}$
$= 6\frac{7}{20}$

$6\frac{3}{8} - 6\frac{7}{20} = \frac{3}{8} - \frac{7}{20}$
$= \frac{15}{40} - \frac{14}{40}$
$= \frac{1}{40}$

Sinister Stan needs $\frac{1}{40}$ ounce of slime.

Problem 4

Gavin had 20 minutes to do a three-problem quiz. He spent $9\frac{3}{4}$ minutes on Problem 1 and $3\frac{4}{5}$ minutes on Problem 2. How much time did he have left for Problem 3? Write the answer in minutes and seconds.

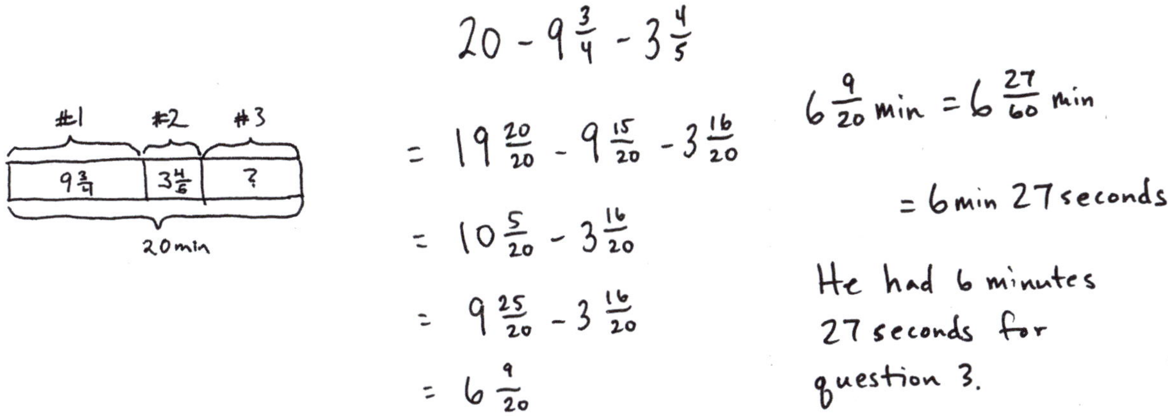

Problem 5

Matt wants to shave $2\frac{1}{2}$ minutes off his 5K race time. After a month of hard training, he managed to lower his overall time from $21\frac{1}{5}$ minutes to $19\frac{1}{4}$ minutes. By how many more minutes does Matt need to lower his race time?

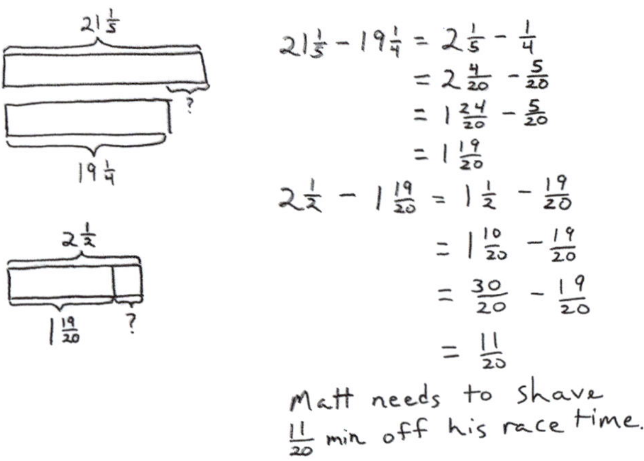

Student Debrief (10 minutes)

Lesson Objective: Solve multi-step word problems; assess reasonableness of solutions using benchmark numbers.

The Student Debrief is intended to invite reflection and active processing of the total lesson experience.

Invite students to review their solutions for the Problem Set. They should check work by comparing answers with a partner before going over answers as a class. Look for misconceptions or misunderstandings that can be addressed in the Debrief. Guide students in a conversation to debrief the Problem Set and process the lesson.

T: Bring your Problem Set to the Debrief. Share, check, and/or explain your answers to your partner.

S: (Work together for 2 minutes.)

T: (Circulate and listen to explanations. Analyze the work to determine which student solutions to display to support the lesson objective.)

T: (Go over answers.) Let's read Problem 4 together, and we'll take a look at 2 different solution strategies.

S: Gavin had 20 minutes to do a three-problem quiz. He spent $9\frac{3}{4}$ minutes on Problem 1 and $3\frac{4}{5}$ minutes on Problem 2. How much time did he have left for Problem 3? Write the answer in minutes and seconds.

T: Discuss what you notice about the two different drawings. (Allow time for students to share.)

T: Jaron, would you share your thinking?

S: The first drawing labeled the whole on the bottom. The second drawing labeled it on the side.

T: How are the drawings similar? Turn and share.

S: (Share.)

T: Keri, what do you think?

S: Both drawings labeled the time for the 3 questions. They also labeled the total amount of time, which is 20 minutes.

T: Let's look at them closely. How did Student A solve the problem? Turn and share.

S: Student A used the total of 20 minutes to subtract the time spent on Problems 1 and 2 to find the leftover time. Then, the student converted $6\frac{9}{20}$ to 6 minutes and 27 seconds.

T: How did Student B solve the problem? Turn and share.

S: Student B converted all the mixed numbers into minutes and seconds. Then, the student used the 20 minutes to subtract the time spent on Problem 1, which is 9 minutes 45 seconds, and the time spent for Problem 2, which is 3 minutes 48 seconds. 6 minutes 27 seconds were left over for Problem 3.

T: Which solution strategy did you like better?

S: The first one. → The first one is a lot shorter than the second one. → The second seems like it should be easy, but it took a long time to write it out with all of the minutes and seconds. → Because it was twentieths, it was really easy to change it to minutes and seconds from $6\frac{9}{20}$ minutes: I just multiplied the fraction by 3 thirds.

Student A's Work

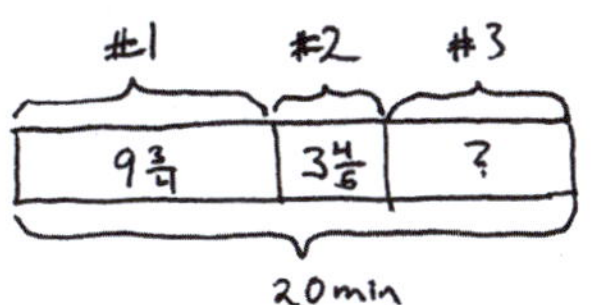

$20 - 9\frac{3}{4} - 3\frac{4}{5}$

$= 19\frac{20}{20} - 9\frac{15}{20} - 3\frac{16}{20}$

$= 10\frac{5}{20} - 3\frac{16}{20}$

$= 9\frac{25}{20} - 3\frac{16}{20}$

$= 6\frac{9}{20}$

$6\frac{9}{20}$ min $= 6\frac{27}{60}$ min

= 6 min 27 seconds

He had 6 minutes 27 seconds for question 3.

Student B's Work

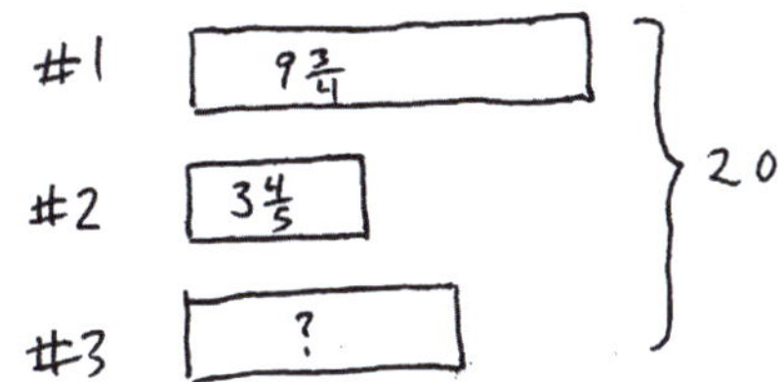

Convert mixed #'s to minutes and seconds.

$9\frac{3}{4}$ min $= 9\frac{45}{60}$ min = 9 min 45 s

$3\frac{4}{5}$ min $= 3\frac{48}{60}$ min = 3 min 48 s

20 min − 9 min 45 s − 3 min 48 s

= 11 min − 45 s − 3 min 48 s

= 10 min 60 s − 45 s − 3 min 48 s

= 10 min 15 s − 3 min 48 s

= 9 min 75 s − 3 min 48 s

= 6 min 27 s

He had 6 min 27 s for question 3.

Optional as time allows: The following is a suggested list of questions to invite reflection and active processing of the total lesson experience. Use those that best support students' ability to articulate the focus of the lesson.

- Did anyone else solve the problem differently? (Students come forward and explain their solution strategies to the class.)
- How did you improve your skills today?

Exit Ticket (3 minutes)

After the Student Debrief, instruct students to complete the Exit Ticket. A review of their work will help with assessing students' understanding of the concepts that were presented in today's lesson and planning more effectively for future lessons. The questions may be read aloud to the students.

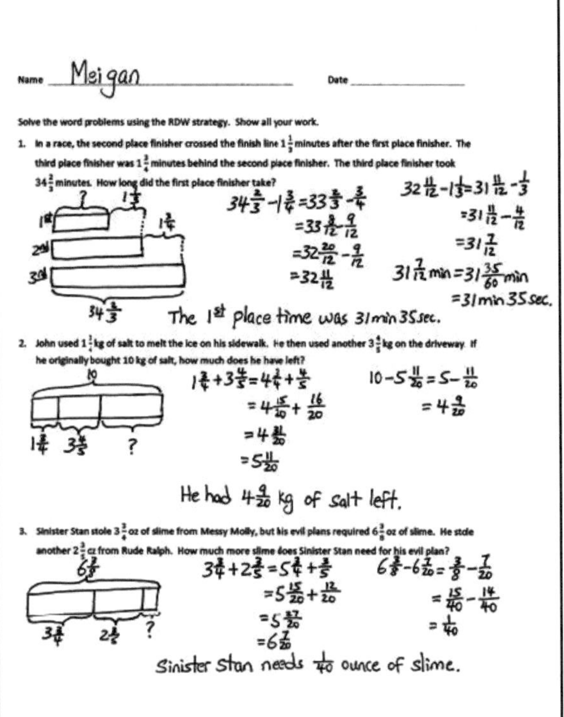

Name Meigan Date

Solve the word problems using the RDW strategy. Show all your work.

1. In a race, the second place finisher crossed the finish line $1\frac{1}{3}$ minutes after the first place finisher. The third place finisher was $1\frac{3}{4}$ minutes behind the second place finisher. The third place finisher took $34\frac{2}{3}$ minutes. How long did the first place finisher take?

$34\frac{2}{3} - 1\frac{3}{4} = 33\frac{2}{3} - \frac{3}{4} = 33\frac{8}{12} - \frac{9}{12} = 32\frac{20}{12} - \frac{9}{12} = 32\frac{11}{12}$

$32\frac{11}{12} - 1\frac{1}{3} = 31\frac{11}{12} - \frac{1}{3} = 31\frac{11}{12} - \frac{4}{12} = 31\frac{7}{12}$

$31\frac{7}{12}$ min $= 31\frac{35}{60}$ min $= 31$ min 35 sec.

The 1st place time was 31 min 35 sec.

2. John used $1\frac{3}{4}$ kg of salt to melt the ice on his sidewalk. He then used another $3\frac{4}{5}$ kg on the driveway. If he originally bought 10 kg of salt, how much does he have left?

$1\frac{3}{4} + 3\frac{4}{5} = 4\frac{3}{4} + \frac{4}{5} = 4\frac{15}{20} + \frac{16}{20} = 4\frac{31}{20} = 5\frac{11}{20}$

$10 - 5\frac{11}{20} = 5 - \frac{11}{20} = 4\frac{9}{20}$

He had $4\frac{9}{20}$ kg of salt left.

3. Sinister Stan stole $3\frac{3}{4}$ oz of slime from Messy Molly, but his evil plans required $6\frac{3}{8}$ oz of slime. He stole another $2\frac{3}{5}$ oz from Rude Ralph. How much more slime does Sinister Stan need for his evil plan?

$3\frac{3}{4} + 2\frac{3}{5} = 5\frac{3}{4} + \frac{3}{5} = 5\frac{15}{20} + \frac{12}{20} = 5\frac{27}{20} = 6\frac{7}{20}$

$6\frac{3}{8} - 6\frac{7}{20} = \frac{3}{8} - \frac{7}{20} = \frac{15}{40} - \frac{14}{40} = \frac{1}{40}$

Sinister Stan needs $\frac{1}{40}$ ounce of slime.

4. Gavin had 20 minutes to do a three-problem quiz. He spent $9\frac{3}{4}$ minutes on question 1 and $3\frac{4}{5}$ minutes on question 2. How much time did he have left for question 3? Write the answer in minutes and seconds.

#1 $9\frac{3}{4}$ | #2 $3\frac{4}{5}$ | #3 ? — 20 min

$20 - 9\frac{3}{4} - 3\frac{4}{5} = 19\frac{20}{20} - 9\frac{15}{20} - 3\frac{16}{20} = 10\frac{5}{20} - 3\frac{16}{20} = 9\frac{25}{20} - 3\frac{16}{20} = 6\frac{9}{20}$

$6\frac{9}{20}$ min $= 6\frac{27}{60}$ min $= 6$ min 27 sec.

He had 6 minutes 27 seconds for question 3.

5. Matt wants to save $2\frac{1}{2}$ minutes on his 5K race time. After a month of hard training he managed to lower his overall time from $21\frac{1}{5}$ minutes to $19\frac{1}{4}$ minutes. By how many more minutes does Matt need to lower his race time?

$21\frac{1}{5} - 19\frac{1}{4} = 2\frac{1}{5} - \frac{1}{4} = 2\frac{4}{20} - \frac{5}{20} = 1\frac{24}{20} - \frac{5}{20} = 1\frac{19}{20}$

$2\frac{1}{2} - 1\frac{19}{20} = 1\frac{1}{2} - \frac{19}{20} = 1\frac{10}{20} - \frac{19}{20} = \frac{30}{20} - \frac{19}{20} = \frac{11}{20}$

Matt needs to save $\frac{11}{20}$ minute off his race time.

A

Number Correct: _______

Circle the Smaller Fraction

1.	$1/2$	$1/4$		23.	$1/4$	$1/8$
2.	$1/2$	$3/4$		24.	$1/4$	$3/8$
3.	$1/2$	$5/8$		25.	$1/4$	$7/12$
4.	$1/2$	$7/8$		26.	$1/4$	$11/12$
5.	$1/2$	$1/10$		27.	$1/6$	$7/12$
6.	$1/2$	$3/10$		28.	$1/6$	$11/12$
7.	$1/2$	$5/12$		29.	$2/3$	$1/6$
8.	$1/2$	$11/12$		30.	$2/3$	$5/6$
9.	$1/2$	$7/10$		31.	$2/3$	$2/9$
10.	$1/5$	$9/10$		32.	$2/3$	$4/9$
11.	$2/5$	$1/10$		33.	$2/3$	$1/12$
12.	$2/5$	$3/10$		34.	$2/3$	$5/12$
13.	$3/5$	$3/10$		35.	$2/3$	$11/12$
14.	$3/5$	$7/10$		36.	$2/3$	$7/12$
15.	$4/5$	$1/10$		37.	$3/4$	$1/8$
16.	$4/5$	$9/10$		38.	$3/4$	$1/8$
17.	$1/3$	$1/9$		39.	$5/6$	$7/12$
18.	$1/3$	$2/9$		40.	$5/6$	$5/12$
19.	$1/3$	$4/9$		41.	$6/7$	$38/42$
20.	$1/3$	$8/9$		42.	$7/8$	$62/72$
21.	$1/3$	$1/12$		43.	$49/54$	$8/9$
22.	$1/3$	$5/12$		44.	$67/72$	$11/12$

B

Number Correct: _______

Improvement: _______

Circle the Smaller Fraction

1.	$^1/_2$	$^1/_6$
2.	$^1/_2$	$^5/_6$
3.	$^1/_2$	$^1/_8$
4.	$^1/_2$	$^3/_8$
5.	$^1/_2$	$^7/_{10}$
6.	$^1/_2$	$^9/_{10}$
7.	$^1/_2$	$^1/_{12}$
8.	$^1/_2$	$^7/_{12}$
9.	$^1/_5$	$^1/_{10}$
10.	$^1/_5$	$^3/_{10}$
11.	$^2/_5$	$^7/_{10}$
12.	$^2/_5$	$^9/_{10}$
13.	$^3/_5$	$^1/_{10}$
14.	$^3/_5$	$^9/_{10}$
15.	$^4/_5$	$^3/_{10}$
16.	$^4/_5$	$^7/_{10}$
17.	$^1/_3$	$^1/_6$
18.	$^1/_3$	$^5/_6$
19.	$^1/_3$	$^5/_9$
20.	$^1/_3$	$^7/_9$
21.	$^1/_3$	$^7/_{12}$
22.	$^1/_3$	$^{11}/_{12}$
23.	$^1/_4$	$^5/_8$
24.	$^1/_4$	$^7/_8$
25.	$^1/_4$	$^1/_{12}$
26.	$^1/_4$	$^5/_{12}$
27.	$^1/_6$	$^1/_{12}$
28.	$^1/_6$	$^5/_{12}$
29.	$^2/_3$	$^1/_9$
30.	$^2/_3$	$^7/_9$
31.	$^2/_3$	$^5/_9$
32.	$^2/_3$	$^8/_9$
33.	$^3/_4$	$^1/_2$
34.	$^3/_4$	$^5/_{12}$
35.	$^3/_4$	$^{11}/_{12}$
36.	$^3/_4$	$^7/_{12}$
37.	$^5/_6$	$^1/_{12}$
38.	$^5/_6$	$^{11}/_{12}$
39.	$^3/_4$	$^5/_8$
40.	$^3/_4$	$^3/_8$
41.	$^6/_7$	$^{34}/_{42}$
42.	$^7/_8$	$^{64}/_{72}$
43.	$^{47}/_{54}$	$^8/_9$
44.	$^{65}/_{72}$	$^{11}/_{12}$

Name ______________________________ Date ____________________

Solve the word problems using the RDW strategy. Show all of your work.

1. In a race, the-second place finisher crossed the finish line $1\frac{1}{3}$ minutes after the winner. The third-place finisher was $1\frac{3}{4}$ minutes behind the second-place finisher. The third-place finisher took $34\frac{2}{3}$ minutes. How long did the winner take?

2. John used $1\frac{3}{4}$ kg of salt to melt the ice on his sidewalk. He then used another $3\frac{4}{5}$ kg on the driveway. If he originally bought 10 kg of salt, how much does he have left?

3. Sinister Stan stole $3\frac{3}{4}$ oz of slime from Messy Molly, but his evil plans require $6\frac{3}{8}$ oz of slime. He stole another $2\frac{3}{5}$ oz of slime from Rude Ralph. How much more slime does Sinister Stan need for his evil plan?

4. Gavin had 20 minutes to do a three-problem quiz. He spent $9\frac{3}{4}$ minutes on Problem 1 and $3\frac{4}{5}$ minutes on Problem 2. How much time did he have left for Problem 3? Write the answer in minutes and seconds.

5. Matt wants to shave $2\frac{1}{2}$ minutes off his 5K race time. After a month of hard training, he managed to lower his overall time from $21\frac{1}{5}$ minutes to $19\frac{1}{4}$ minutes. By how many more minutes does Matt need to lower his race time?

Name ______________________________ Date ______________

Solve the word problem using the RDW strategy. Show all of your work.

Cheryl bought a sandwich for $5\frac{1}{2}$ dollars and a drink for \$2.60. If she paid for her meal with a \$10 bill, how much money did she have left? Write your answer as a fraction and in dollars and cents.

Name ______________________________ Date ______________

Solve the word problems using the RDW strategy. Show all of your work.

1. A baker buys a 5 lb bag of sugar. She uses $1\frac{2}{3}$ lb to make some muffins and $2\frac{3}{4}$ lb to make a cake. How much sugar does she have left?

2. A boxer needs to lose $3\frac{1}{2}$ kg in a month to be able to compete as a flyweight. In three weeks, he lowers his weight from 55.5 kg to 53.8 kg. How many kilograms must the boxer lose in the final week to be able to compete as a flyweight?

3. A construction company builds a new rail line from Town A to Town B. They complete $1\frac{1}{4}$ miles in their first week of work and $1\frac{2}{3}$ miles in the second week. If they still have $25\frac{3}{4}$ miles left to build, what is the distance from Town A to Town B?

4. A catering company needs 8.75 lb of shrimp for a small party. They buy $3\frac{2}{3}$ lb of jumbo shrimp, $2\frac{5}{8}$ lb of medium-sized shrimp, and some mini-shrimp. How many pounds of mini-shrimp do they buy?

5. Mark breaks up a 9-hour drive into 3 segments. He drives $2\frac{1}{2}$ hours before stopping for lunch. After driving some more, he stops for gas. If the second segment of his drive was $1\frac{2}{3}$ hours longer than the first segment, how long did he drive after stopping for gas?

Lesson 16

Objective: Explore part-to-whole relationships.

Suggested Lesson Structure

- Fluency Practice (15 minutes)
- Concept Development (35 minutes)
- Student Debrief (10 minutes)

Total Time (60 minutes)

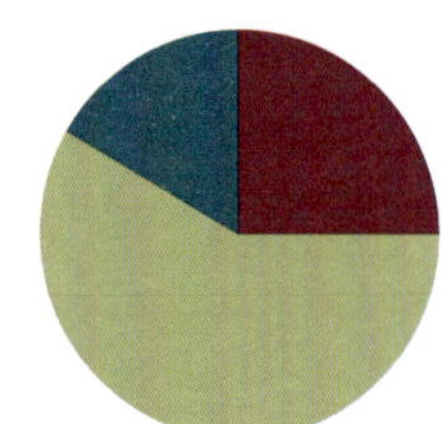

Fluency Practice (15 minutes)

- Break Apart the Whole **4.NF.3c** (5 minutes)
- Make a Like Unit **5.NF.1** (5 minutes)
- Add Fractions with Answers Greater than 1 **5.NF.1** (5 minutes)

Break Apart the Whole (5 minutes)

Materials: (S) Personal white board

Note: Students decompose fractions greater than 1 (i.e., improper fractions) into a whole and a fraction in preparation for today's Concept Development.

T: I'll give you a fraction greater than one, and you'll break out the whole by writing the addition fraction sentence. For example, I say $\frac{3}{2}$, and you write $1 + \frac{1}{2}$. (The teacher can also ask students to write out the whole number fraction plus fraction, i.e., $\frac{2}{2} + \frac{1}{2}$.)

T: $\frac{4}{3}$.

S: (Write $1 + \frac{1}{3}$.)

T: $\frac{7}{5}$.

S: (Write $1 + \frac{2}{5}$.)

T: $\frac{19}{17}$.

S: (Write $1 + \frac{2}{17}$.)

T: $\frac{13}{3}$.

S: (Write $4 + \frac{1}{3}$.)

T: $\frac{31}{6}$.

S: (Write $5 + \frac{1}{6}$.)

T: (Continue with a sequence appropriate for students.) Share with your partner. What's your strategy of breaking out the whole?

S: (Share.)

T: Excellent!

Make a Like Unit (5 minutes)

Materials: (S) Personal white board

Note: Students make like units, which is a prerequisite skill for advanced work with fractions.

T: What does *like unit* mean?

S: When you add or subtract fractions, if the denominators are the same, then they are like units.

T: Tell your partner how we find like units.

S: (Share.)

T: I'll say two numbers. You make a like unit and write it on your personal white board.

T: 3 and 2.

S: (Write and show 6.)

Continue with the following possible sequence: 4 and 3; 2 and 4; 2 and 6; 3 and 9; 3 and 12; 4 and 8; 6 and 8.

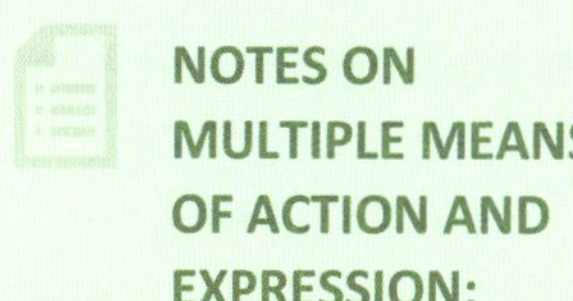

NOTES ON MULTIPLE MEANS OF ACTION AND EXPRESSION:

Some students may provide the smallest like unit; others may not. Accept a range of answers. Notice which students consistently do not show the smallest or easiest like unit. It may be that they need extra support.

Add Fractions with Answers Greater than 1 (5 minutes)

Materials: (S) Personal white board

Note: Students recognize and analyze fractions greater than 1 in preparation for today's problem-solving set.

T: I'll say an equation. You write and solve it. If the answer is greater than 1, put a dot next to it. Leave room to write all of the equations on your personal white board without erasing.

T: $\frac{3}{3} + \frac{1}{3} =$ ____.

S: (Show $\frac{3}{3} + \frac{1}{3} = \frac{4}{3}$ •)

T: $\frac{2}{2} + \frac{3}{2} =$ ____.

S: (Show $\frac{2}{2} + \frac{3}{2} = \frac{5}{2}$ •)

NOTES ON MULTIPLE MEANS OF ACTION AND EXPRESSION:

Depending on the class, make the activity slightly more complex by eliminating the personal white board. Have students complete the activity orally.

T: $\frac{2}{4} + \frac{1}{4} =$ ____.

S: (Show $\frac{2}{4} + \frac{1}{4} = \frac{3}{4}$.)

T: (Continue, alternating between equations that have answers greater than 1 or 2 with those that don't.) What is different about answers that are greater than 1 and those that are less?

S: Answers greater than 1 all have a numerator that is greater than the denominator.

T: Some of these answers are greater than 2. Circle those.

S: (Find and circle the appropriate answers.)

T: Talk to your partner about the difference between answers that are greater than 1 and those that are greater than 2.

S: Numerators are greater than denominators in both. → Yes, but greater than 2 means the denominator has to fit inside the numerator at least twice, too.

Concept Development (35 minutes)

Materials: (S) Problem Set

T: Today, you are going to work in pairs to solve some ribbon and wire problems. I am going to be an observer for the most part, just listening and watching until the Debrief. You have 30 minutes to reason about and solve 3 problems. I will let you know when you have 10 minutes and then 5 minutes remaining. You can use any materials in the classroom, but I ask that you work only with your partner. The work will be scored with a rubric. Each question can earn 4 points.

- Question 1: Each correct answer including the drawing is 1 point.
- Questions 2 and 3: Clear drawing: 1 point. Labeled drawing: 1 point. Correct equation and answer: 1 point. Correct statement of your answer: 1 point. The total possible number of points is 12.

1. Draw the following ribbons. When finished, compare your work to your partner's.

   a. 1 ribbon. The piece shown below is only $\frac{1}{3}$ of the whole. Complete the drawing to show the whole ribbon.

   b. 1 ribbon. The piece shown below is $\frac{4}{5}$ of the whole. Complete the drawing to show the whole ribbon.

c. 2 ribbons, A and B. One third of A is equal to all of B. Draw a picture of the ribbons.

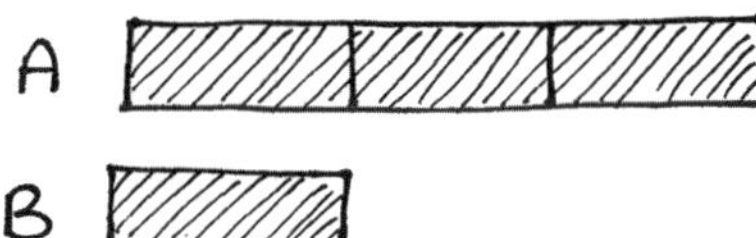

d. 3 ribbons, C, D, and E. C is half the length of D. E is twice as long as D. Draw a picture of the ribbons.

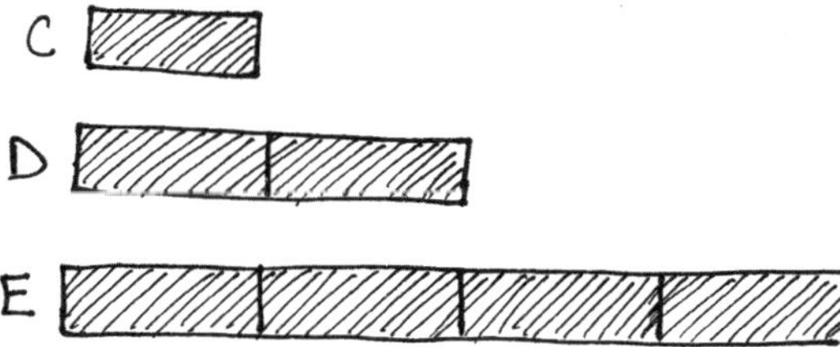

2. Half of Robert's piece of wire is equal to $\frac{2}{3}$ of Maria's wire. The total length of their wires is 10 feet. How much longer is Robert's wire than Maria's?

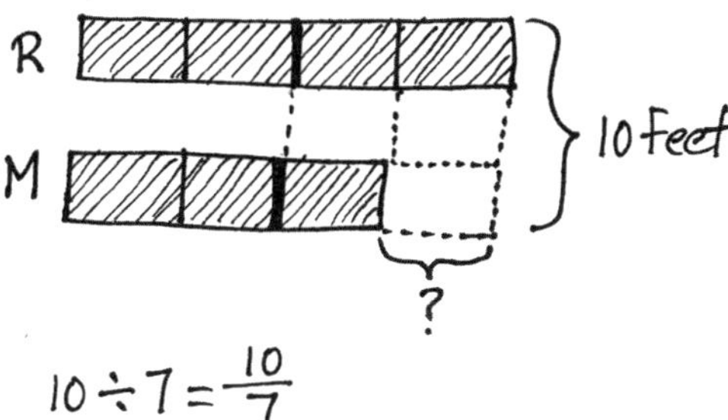

Robert's wire is $\frac{10}{7}$ feet longer than Maria's.

3. Half of Sarah's wire is equal to $\frac{2}{5}$ of Daniel's. Chris has 3 times as much as Sarah. In all, their wire measures 6 ft. How long is Sarah's wire in feet?

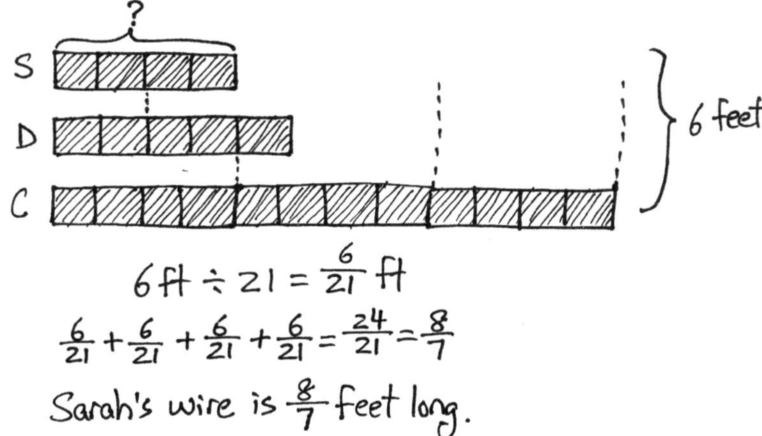

MP.1

This lesson is an opportunity for students to *make sense of problems and persevere in solving them* (MP.1). Observe with as little interference as possible. For students who have language barriers, support by pairing appropriately for primary language, or provide a translation of the problem. As students work, circulate and make decisions about which work to share with the class and in what order.

Student Debrief (10 minutes)

Lesson Objective: Explore part-to-whole relationships.

The Student Debrief is intended to invite reflection and active processing of the total lesson experience.

Invite students to review their solutions for the Problem Set. They should check work by comparing answers with a partner before going over answers as a class. Look for misconceptions or misunderstandings that can be addressed in the Debrief. Guide students in a conversation to debrief the Problem Set and process the lesson.

T: Let's work together now to analyze your solutions. Compare your solutions from the first page with the solutions of your neighbor to the right.

T: What surprises did you have on this first page?

S: We don't usually think about a shape being a fraction of something. We usually fold it or partition it to make a fraction, so it was new to think of something as a fraction.

T: (After reviewing work from the first page.) Let's analyze John and Erica's work from the two story problems (Students' Work 1). Take a moment to talk to your partner about precisely what you see when you look at their work and how that relates to the questions.

T: (After students have analyzed the work.) Does anyone have a question for John and Erica?

S: Can you explain how you got 21 units in Problem 3?

S: (Authors explain.)

Students' Work 1: John and Erica

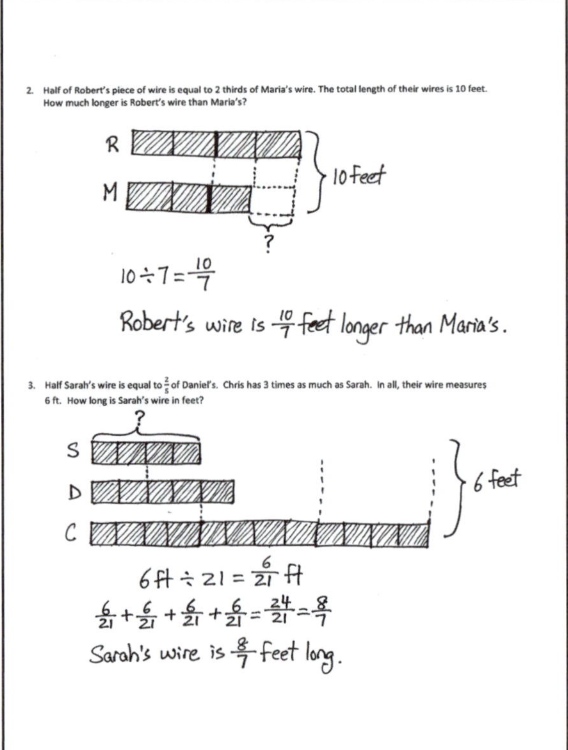

Students' work #2: Jacqueline and Perry

2. Half of Robert's piece of wire is equal to 2 thirds of Maria's wire. The total length of their wires is 10 feet. How much longer is Robert's wire than Maria's?

R:
M:
10 ft
?

7 units = 10 ft
1 unit = $\frac{10}{7}$ ft

Robert's wire is $\frac{10}{7}$ ft longer than Maria's.

3. Half of Sarah's wire is equal to $\frac{2}{5}$ of Daniel's. Chris has 3 times as much as Sarah. In all, their wire measures 6 ft. How long is Sarah's wire in feet?

S:
D:
C:
6 ft

21 units = 6 ft
1 unit = $\frac{6}{21}$ ft = $\frac{2}{7}$ ft
4 units = $4 \times \frac{2}{7}$ ft = $\frac{8}{7}$ ft = $1\frac{1}{7}$ ft

Sarah's wire is $1\frac{1}{7}$ ft long.

T: Retell the explanation to your partner in your own words.

T: (After time for conversation.) Let's compare their work with Jacqueline and Perry's (Students' Work 2). First, analyze the new team's work by itself for a minute.

T: (After analysis and questioning the authors.)

T: Now, let's compare these two pieces of student work. What is the same, and what is different?

S: (Discuss in pairs and at times as a whole group.)

T: Be sure to compare their work numerically, too. What number sentences did they use? How do their number sentences relate to each other's work? For example, where do we see $4 \times \frac{2}{7}$ in John and Erica's work?

Once the analysis is complete, encourage students to score their own work according to the predetermined rubric. This task brings students back to a fraction as a quotient and a quotient as a fraction. It also heightens their fraction number sense right as they are about to embark on the multiplication and division module.

Exit Ticket (3 minutes)

After the Student Debrief, instruct students to complete the Exit Ticket. A review of their work will help with assessing students' understanding of the concepts that were presented in today's lesson and planning more effectively for future lessons. The questions may be read aloud to the students.

Names ______________________________ and ________________________ Date ___________________

1. Draw the following ribbons. When finished, compare your work to your partner's.

   a. 1 ribbon. The piece shown below is only $\frac{1}{3}$ of the whole. Complete the drawing to show the whole ribbon.

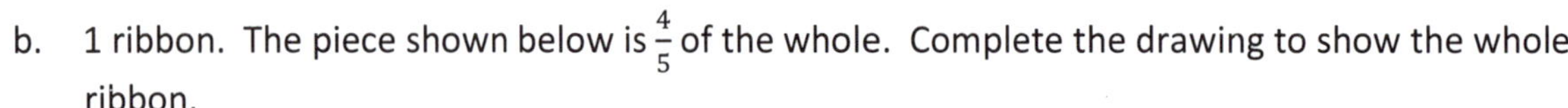

   b. 1 ribbon. The piece shown below is $\frac{4}{5}$ of the whole. Complete the drawing to show the whole ribbon.

   c. 2 ribbons, A and B. One third of A is equal to all of B. Draw a picture of the ribbons.

   d. 3 ribbons, C, D, and E. C is half the length of D. E is twice as long as D. Draw a picture of the ribbons.

2. Half of Robert's piece of wire is equal to $\frac{2}{3}$ of Maria's wire. The total length of their wires is 10 feet. How much longer is Robert's wire than Maria's?

3. Half of Sarah's wire is equal to $\frac{2}{5}$ of Daniel's. Chris has 3 times as much as Sarah. In all, their wire measures 6 ft. How long is Sarah's wire in feet?

Name ____ Date ______________________

Draw the following ribbons.

a. 1 ribbon. The piece shown below is only $\frac{2}{3}$ of the whole. Complete the drawing to show the whole ribbon.

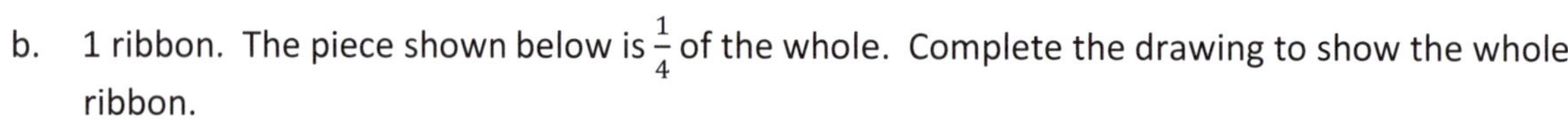

b. 1 ribbon. The piece shown below is $\frac{1}{4}$ of the whole. Complete the drawing to show the whole ribbon.

c. 3 ribbons, A, B, and C. 1 third of A is the same length as B. C is half as long as B. Draw a picture of the ribbons.

Name ______________________________ Date ______________

Draw the following roads.

a. 1 road. The piece shown below is only $\frac{3}{7}$ of the whole. Complete the drawing to show the whole road.

b. 1 road. The piece shown below is $\frac{1}{6}$ of the whole. Complete the drawing to show the whole road.

c. 3 roads, A, B, and C. B is three times longer than A. C is twice as long as B. Draw the roads. What fraction of the total length of the roads is the length of A? If Road B is 7 miles longer than Road A, what is the length of Road C?

d. Write your own road problem with 2 or 3 lengths.

Name ______________________________ Date ________________

1. On Sunday, Sheldon bought $4\frac{1}{2}$ kg of plant food. He used $1\frac{2}{3}$ kg on his strawberry plants and used $\frac{1}{4}$ kg for his tomato plants.

   a. How many kilograms of plant food did Sheldon have left? Write one or more equations to show how you reached your answer.

   b. Sheldon wants to feed his strawberry plants 2 more times and his tomato plants one more time. He will use the same amounts of plant food as before. How much plant food will he need? Does he have enough left to do so? Explain your answer using words, pictures, or numbers.

2. Sheldon harvests the strawberries and tomatoes in his garden.

   a. He picks $1\frac{2}{5}$ kg less strawberries in the morning than in the afternoon. If Sheldon picks $2\frac{1}{4}$ kg in the morning, how many kilograms of strawberries does he pick in the afternoon? Explain your answer using words, pictures, or equations.

   b. Sheldon also picks tomatoes from his garden. He picked $5\frac{3}{10}$ kg, but 1.5 kg were rotten and had to be thrown away. How many kilograms of tomatoes were not rotten? Write an equation that shows how you reached your answer.

   c. After throwing away the rotten tomatoes, did Sheldon get more kilograms of strawberries or tomatoes? How many more kilograms? Explain your answer using an equation.

End-of-Module Assessment Task Standards Addressed	Topics A–D

Use equivalent fractions as a strategy to add and subtract fractions.

5.NF.1 Add and subtract fractions with unlike denominators (including mixed numbers) by replacing given fractions with equivalent fractions in such a way as to produce an equivalent sum or difference of fractions with like denominators. *For example, 2/3 + 5/4 = 8/12 + 15/12 = 23/12. (In general, a/b + c/d = (ad + bc)/bd.)*

5.NF.2 Solve word problems involving addition and subtraction of fractions referring to the same whole, including cases of unlike denominators, e.g., by using visual fraction models or equations to represent the problem. Use benchmark fractions and number sense of fractions to estimate mentally and assess the reasonableness of answers. *For example, recognize an incorrect result 2/5 + 1/2 = 3/7, by observing that 3/7 < 1/2.*

Evaluating Student Learning Outcomes

A Progression Toward Mastery is provided to describe steps that illuminate the gradually increasing understandings that students develop on their way to proficiency. In this chart, this progress is presented from left (Step 1) to right (Step 4). The learning goal for students is to achieve Step 4 mastery. These steps are meant to help teachers and students identify and celebrate what the students CAN do now and what they need to work on next.

A Progression Toward Mastery				
Assessment Task Item and Standards Assessed	**STEP 1 Little evidence of reasoning without a correct answer. (1 Point)**	**STEP 2 Evidence of some reasoning without a correct answer. (2 Points)**	**STEP 3 Evidence of some reasoning with a correct answer or evidence of solid reasoning with an incorrect answer. (3 Points)**	**STEP 4 Evidence of solid reasoning with a correct answer. (4 Points)**
1(a) 5.NF.1 5.NF.2	The work shows little evidence of conceptual or procedural strength.	Student obtains an incorrect answer and has trouble manipulating the units or setting up the problem.	Student obtains the correct answer but does not show an equation or does not obtain the correct answer through a very small calculation error. The part–whole thinking is completely accurate.	Student displays complete confidence in applying part–whole thinking to a word problem with fractions, giving the correct answer of $2\frac{14}{24}$ kg or $2\frac{7}{12}$ kg.
1(b) 5.NF.1 5.NF.2	Student was unable to make sense of the problem in any intelligible way.	Student's solution is incorrect and, though showing signs of real thought, is not developed or does not connect to the story's situation.	Student has the correct answer to the first question but fails to answer the second question. OR Student has reasoned through the problem well, setting up the equation correctly, but making a careless error.	Student correctly: ▪ Calculates that Sheldon needs $3\frac{7}{12}$ kg of plant food. ▪ Notes that $3\frac{7}{12}$ kg is more than $2\frac{7}{12}$ kg, so Sheldon does not have enough plant food.
2(a) 5.NF.1 5.NF.2	The solution is incorrect and shows little evidence of understanding of the need for like units.	Student shows evidence of beginning to understand adding fractions with unlike denominators but cannot apply that knowledge to this part–whole comparison.	Student calculates correctly and sets up the part–whole situation correctly but fails to write a complete statement. OR Student fully answers the question but makes one small calculation error that is clearly careless, such as copying a number wrong.	Student is able to apply part–whole thinking to correctly answer $3\frac{13}{20}$ kg and explains the answer using words, pictures, or numbers.

A Progression Toward Mastery				
2(b) **5.NF.1** **5.NF.2**	The solution is incorrect and shows no evidence of being able to work with decimals and fractions simultaneously.	Student shows evidence of recognizing how to convert fractions to decimals or decimals to fractions but fails to do so correctly.	Student calculates correctly but may be less than perfectly clear in stating his solution. For example, "The answer is $3\frac{4}{5}$," is not a clearly stated solution.	Student gives a correct equation and correct answer of $3\frac{8}{10}$ kg or $3\frac{4}{5}$ kg and explains the answer using words, pictures, or numbers.
2(c) **5.NF.1** **5.NF.2**	The solution is incorrect and shows little evidence of understanding of fraction comparison.	Student may have compared correctly but calculated incorrectly and/or does not explain the meaning of her numerical solution in the context of the story.	Student may have compared correctly but calculated incorrectly and/or does not explain the meaning of her numerical solution in the context of the story.	Student correctly: ▪ Responds that the garden produced more strawberries. ▪ Responds that there were $2\frac{1}{10}$ kg or 2.1 kg more strawberries. ▪ Gives an equation such as $5\frac{9}{10} - 3\frac{8}{10} = 2\frac{9}{10} - \frac{8}{10} = 2\frac{1}{10}$.

Name Jacqueline Date ____________

1) On Sunday, Sheldon bought $4\frac{1}{2}$ kg of plant food. He used $1\frac{2}{3}$ kg on his strawberry plants, and used $\frac{1}{4}$ kg for his tomato plants.

a) How many kilograms of plant food did Sheldon have left? Write one or more equations to show how you reached your answer.

$4\frac{1}{2}$ kg

left | $\frac{1}{4}$ kg | $1\frac{2}{3}$ kg

?

$4\frac{1}{4} - 1\frac{2}{3} = 3\frac{1}{4} - \frac{2}{3}$

$= 3\frac{3}{12} - \frac{8}{12}$

$= 2\frac{15}{12} - \frac{8}{12}$

$= 2\frac{7}{12}$

Sheldon had $2\frac{7}{12}$ kg left.

b) Sheldon wants to feed his strawberry plants 2 more times, and his tomato plants one more time. He will use the same amounts of plant food as before. How much plant food will he need? Does he have enough left to do so? Explain your answer using words, pictures or numbers.

$1\frac{2}{3} + 1\frac{2}{3} = 2\frac{2}{3} + \frac{2}{3}$

$= 3\frac{1}{3}$

$3\frac{1}{3} + \frac{1}{4} = 3\frac{4}{12} + \frac{3}{12}$

$= 3\frac{7}{12}$

No, Sheldon does not have enough because.

$2\frac{7}{12} < 3\frac{7}{12}$.

↓ what he has left

↓ what he needs.

2) Sheldon harvests the strawberries and tomatoes in his garden.

a. He picks $1\frac{2}{5}$ kg less strawberries in the morning than in the afternoon. If Sheldon picks $2\frac{1}{4}$ kg in the morning, how many kilograms of strawberries does he pick in the afternoon? Explain your answer using words, pictures or equations.

M [$2\frac{1}{4}$ kg]

A [| $1\frac{2}{5}$]

?

$$2\frac{1}{4} + 1\frac{2}{5} = 3\frac{1}{4} + \frac{2}{5}$$
$$= 3\frac{5}{20} + \frac{8}{20}$$
$$= 3\frac{13}{20}$$

Sheldon picked $3\frac{13}{20}$ kg strawberries in the afternoon.

b) Sheldon also picks tomatoes from his garden. He picked $5\frac{3}{10}$ kg but 1.5 kg were rotten and had to be thrown away. How many kilograms of tomatoes were not rotten? Write an equation that shows how you reached your answer.

$$5\frac{3}{10} - 1\frac{5}{10} = 4\frac{3}{10} - \frac{5}{10}$$
$$= 3\frac{13}{10} - \frac{5}{10}$$
$$= 3\frac{8}{10}$$

$3\frac{8}{10}$ kg or $3\frac{4}{5}$ kg were not rotten.

c) After throwing away the rotten tomatoes, did Sheldon get more kilograms of strawberries or tomatoes? How many more kilograms? Explain your answer using an equation.

Tomatoes: $3\frac{8}{10}$ kg

Strawberries: $2\frac{1}{4}$ kg $+ 2\frac{1}{4}$ kg $+ 1\frac{2}{5}$ kg

$= 4\frac{1}{2}$ kg $+ 1\frac{2}{5}$ kg

$= 4\frac{5}{10}$ kg $+ 1\frac{4}{10}$ kg

$= 5\frac{9}{10}$ kg

$5\frac{9}{10}$ kg $> 3\frac{8}{10}$ kg

Sheldon got more strawberries than tomatoes.

$5\frac{9}{10} - 3\frac{8}{10} = 2\frac{1}{10}$

He got $2\frac{1}{10}$ kg more strawberries.

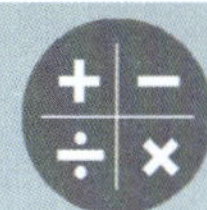

Answer Key

GRADE 5 • MODULE 3

Addition and Subtraction of Fractions

Lesson 1

Sprint

Side A

1.	2	12.	2	23.	4	34.	7
2.	5	13.	2	24.	7	35.	4
3.	2	14.	2	25.	7	36.	8
4.	3	15.	2	26.	7	37.	4
5.	3	16.	3	27.	4	38.	7
6.	2	17.	3	28.	4	39.	8
7.	2	18.	3	29.	6	40.	4
8.	4	19.	4	30.	6	41.	9
9.	3	20.	4	31.	6	42.	9
10.	3	21.	3	32.	9	43.	9
11.	2	22.	3	33.	8	44.	3

Side B

1.	3	12.	3	23.	7	34.	6
2.	2	13.	2	24.	7	35.	3
3.	3	14.	3	25.	7	36.	4
4.	2	15.	3	26.	7	37.	9
5.	2	16.	3	27.	9	38.	8
6.	5	17.	4	28.	9	39.	9
7.	2	18.	4	29.	6	40.	12
8.	2	19.	2	30.	6	41.	8
9.	2	20.	2	31.	6	42.	9
10.	4	21.	4	32.	8	43.	12
11.	3	22.	4	33.	9	44.	4

Problem Set

1. Number line marked 0 and 1 on top and $\frac{0}{2}, \frac{1}{2}, \frac{2}{2}$ on the bottom; $\frac{2}{4}$ shaded; $\frac{3}{6}$ shaded, $\frac{1}{2} = \frac{1 \times 3}{2 \times 3} = \frac{3}{6}$; $\frac{4}{8}$ shaded, $\frac{1}{2} = \frac{1 \times 4}{2 \times 4} = \frac{4}{8}$; $\frac{5}{10}$ shaded, $\frac{1}{2} = \frac{1 \times 5}{2 \times 5} = \frac{5}{10}$
2. Number line marked 0 and 1 on top and $\frac{0}{3}, \frac{1}{3}, \frac{2}{3}, \frac{3}{3}$, on the bottom; $\frac{2}{6}$ shaded, $\frac{1}{3} = \frac{1 \times 2}{3 \times 2} = \frac{2}{6}$; $\frac{3}{9}$ shaded, $\frac{1}{3} = \frac{1 \times 3}{3 \times 3} = \frac{3}{9}$; $\frac{4}{12}$ shaded, $\frac{1}{3} = \frac{1 \times 4}{3 \times 4} = \frac{4}{12}$; $\frac{5}{15}$ shaded, $\frac{1}{3} = \frac{1 \times 5}{3 \times 5} = \frac{5}{15}$
3. Number line marked 0 and 1 on top and $\frac{0}{4}, \frac{1}{4}, \frac{2}{4}, \frac{3}{4}, \frac{4}{4}$, on the bottom; $\frac{6}{8}$ shaded, $\frac{3}{4} = \frac{3 \times 2}{4 \times 2} = \frac{6}{8}$; $\frac{9}{12}$ shaded, $\frac{3}{4} = \frac{3 \times 3}{4 \times 3} = \frac{9}{12}$; $\frac{12}{16}$ shaded, $\frac{3}{4} = \frac{3 \times 4}{4 \times 4} = \frac{12}{16}$; $\frac{15}{20}$ shaded, $\frac{3}{4} = \frac{3 \times 5}{4 \times 5} = \frac{15}{20}$
4. Number line marked 0 and 1 on top and $\frac{0}{5}, \frac{1}{5}, \frac{2}{5}, \frac{3}{5}, \frac{4}{5}, \frac{5}{5}, \frac{6}{5}, \frac{7}{5}$ on the bottom; examples may vary.

Exit Ticket

Number line marked 0 and 1 on top and $\frac{0}{6}, \frac{1}{6}, \frac{2}{6}, \frac{3}{6}, \frac{4}{6}, \frac{5}{6}, \frac{6}{6}$ on the bottom; $\frac{2}{12}$ shaded; $\frac{3}{18}$ shaded, $\frac{1}{6} = \frac{1 \times 3}{6 \times 3} = \frac{3}{18}$; $\frac{4}{24}$ shaded, $\frac{1}{6} = \frac{1 \times 4}{6 \times 4} = \frac{4}{24}$

Homework

1. Number line marked 0 and 1 on top and $\frac{0}{3}, \frac{1}{3}, \frac{2}{3}, \frac{3}{3}$ on the bottom; $\frac{2}{6}$ shaded; $\frac{3}{9}$ shaded, $\frac{1}{3} = \frac{1 \times 3}{3 \times 3} = \frac{3}{9}$; $\frac{4}{12}$ shaded, $\frac{1}{3} = \frac{1 \times 4}{3 \times 4} = \frac{4}{12}$; $\frac{5}{15}$ shaded, $\frac{1}{3} = \frac{1 \times 5}{3 \times 5} = \frac{5}{15}$
2. Number line marked 0 and 1 on top and $\frac{0}{4}, \frac{1}{4}, \frac{2}{4}, \frac{3}{4}, \frac{4}{4}$ on the bottom; $\frac{2}{8}$ shaded, $\frac{1}{4} = \frac{1 \times 2}{4 \times 2} = \frac{2}{8}$; $\frac{3}{12}$ shaded, $\frac{1}{4} = \frac{1 \times 3}{4 \times 3} = \frac{3}{12}$; $\frac{4}{16}$ shaded, $\frac{1}{4} = \frac{1 \times 4}{4 \times 4} = \frac{4}{16}$; $\frac{5}{20}$ shaded, $\frac{1}{4} = \frac{1 \times 5}{4 \times 5} = \frac{5}{20}$
3. Number line marked 0 and 1 on top and $\frac{0}{5}, \frac{1}{5}, \frac{2}{5}, \frac{3}{5}, \frac{4}{5}, \frac{5}{5}$, on the bottom; $\frac{8}{10}$ shaded, $\frac{4}{5} = \frac{4 \times 2}{5 \times 2} = \frac{8}{10}$; $\frac{12}{15}$ shaded, $\frac{4}{5} = \frac{4 \times 3}{5 \times 3} = \frac{12}{15}$; $\frac{16}{20}$ shaded, $\frac{4}{5} = \frac{4 \times 4}{5 \times 4} = \frac{16}{20}$; $\frac{20}{25}$ shaded, $\frac{4}{5} = \frac{4 \times 5}{5 \times 5} = \frac{20}{25}$
4. Number line marked 0 and 1 on top and $\frac{0}{8}, \frac{1}{8}, \frac{2}{8}, \frac{3}{8}, \frac{4}{8}, \frac{5}{8}, \frac{6}{8}, \frac{7}{8}, \frac{8}{8}, \frac{9}{8}$ on the bottom; examples may vary.

Lesson 2

Sprint

Side A

1. 2	12. 12	23. 4	34. 40
2. 10	13. 12	24. 8	35. 54
3. 4	14. 1	25. 2	36. 5
4. 6	15. 3	26. 4	37. 56
5. 8	16. 5	27. 15	38. 4
6. 6	17. 2	28. 35	39. 7
7. 4	18. 4	29. 4	40. 9
8. 9	19. 1	30. 4	41. 72
9. 6	20. 2	31. 42	42. 84
10. 2	21. 4	32. 8	43. 5
11. 6	22. 3	33. 9	44. 6

Side B

1. 10	12. 3	23. 12	34. 64
2. 4	13. 9	24. 12	35. 45
3. 6	14. 2	25. 3	36. 9
4. 8	15. 1	26. 3	37. 48
5. 4	16. 1	27. 25	38. 9
6. 2	17. 5	28. 28	39. 1
7. 6	18. 5	29. 3	40. 5
8. 3	19. 3	30. 5	41. 12
9. 9	20. 3	31. 35	42. 36
10. 8	21. 3	32. 9	43. 4
11. 8	22. 4	33. 8	44. 7

Problem Set

1. a. $\frac{3}{5}$; number line drawn
   b. $\frac{3}{3}$ or 1; number line drawn
   c. $\frac{9}{10}$; number line drawn
   d. $1\frac{3}{4}$; number line drawn
2. a. Answers will vary; number line drawn
   b. Answers will vary.
   c. Answers will vary.
   d. Answers will vary.
3. a. $1\frac{2}{7}$
   b. $4\frac{1}{2}$
   c. $4\frac{4}{7}$; number line drawn
   d. $2\frac{6}{9}$; number line drawn
4. $2\frac{4}{8}$ yd or $2\frac{1}{2}$ yd; number line drawn

Exit Ticket

1. a. $\frac{7}{5}$; number line drawn
   b. $\frac{8}{3}$; number line drawn
2. a. Answers will vary.
   b. Answers will vary; number line drawn

Homework

1. a. $\frac{5}{9}$; number line drawn
   b. $\frac{4}{4}$; number line drawn
   c. $\frac{6}{7}$; number line drawn
   d. $\frac{7}{5}$; number line drawn
2. a. Answers will vary; number line drawn
   b. Answers will vary.
   c. Answers will vary.
   d. Answers will vary.
3. a. $1\frac{4}{5}$
   b. $3\frac{1}{2}$
   c. $3\frac{4}{7}$; number line drawn
   d. $2\frac{3}{9}$ or $2\frac{1}{3}$; number line drawn
4. $4\frac{5}{10}$ m or $4\frac{1}{2}$ m; number line drawn

Lesson 3

Sprint

Side A

1.	2	12.	12	23.	4	34.	40
2.	10	13.	12	24.	8	35.	54
3.	4	14.	1	25.	2	36.	5
4.	6	15.	3	26.	4	37.	56
5.	8	16.	5	27.	15	38.	4
6.	6	17.	2	28.	35	39.	7
7.	4	18.	4	29.	4	40.	9
8.	9	19.	1	30.	4	41.	72
9.	6	20.	2	31.	42	42.	84
10.	2	21.	4	32.	8	43.	5
11.	6	22.	3	33.	9	44.	6

Side B

1.	10	12.	3	23.	12	34.	64
2.	4	13.	9	24.	12	35.	45
3.	6	14.	2	25.	3	36.	9
4.	8	15.	1	26.	3	37.	48
5.	4	16.	1	27.	25	38.	9
6.	2	17.	5	28.	28	39.	1
7.	6	18.	5	29.	3	40.	5
8.	3	19.	3	30.	5	41.	12
9.	9	20.	3	31.	35	42.	36
10.	8	21.	3	32.	9	43.	4
11.	8	22.	4	33.	8	44.	7

Problem Set

1. a. Accurate picture drawn; $\frac{5}{6}$
   b. Accurate picture drawn; $\frac{8}{15}$
   c. Accurate picture drawn; $\frac{7}{12}$
   d. Accurate picture drawn; $\frac{10}{21}$
   e. Accurate picture drawn; $\frac{19}{20}$
   f. Accurate picture drawn; $\frac{20}{21}$
2. Accurate picture and number sentence shown; $\frac{1}{2}$ yd
3. Accurate picture and number sentence shown; $\frac{11}{12}$ qt
4. Accurate picture and number sentence shown; $\frac{13}{20}$; $\frac{7}{20}$

Exit Ticket

1. Accurate picture drawn; $\frac{7}{10}$
2. $\frac{13}{20}$ hr; 39 min

Homework

1. a. Accurate picture drawn; $\frac{7}{12}$
   b. Accurate picture drawn; $\frac{9}{20}$
   c. Accurate picture drawn; $\frac{5}{12}$
   d. Accurate picture drawn; $\frac{14}{45}$
   e. Accurate picture drawn; $\frac{13}{20}$
   f. Accurate picture drawn; $1\frac{1}{35}$
2. Accurate picture and number sentence shown; $\frac{11}{12}$ mi
3. Accurate picture and number sentence shown; $\frac{19}{24}$; $\frac{5}{24}$
4. Accurate picture and number sentence shown; $\frac{17}{30}$; $\frac{13}{30}$

Lesson 4

Problem Set

1. a. Accurate picture drawn; $1\frac{1}{6}$
   b. Accurate picture drawn; $1\frac{5}{12}$
   c. Accurate picture drawn; $1\frac{1}{10}$
   d. Accurate picture drawn; $1\frac{3}{14}$
   e. Accurate picture drawn; $1\frac{7}{12}$
   f. Accurate picture drawn; $1\frac{2}{21}$

2. Accurate picture and number sentence shown; $1\frac{3}{20}$ lb

3. Accurate picture and number sentence shown; $1\frac{9}{20}$ hr; $\frac{11}{20}$ hr

Exit Ticket

1. Accurate picture drawn; $1\frac{1}{12}$

2. $1\frac{11}{20}$ L

Homework

1. a. Accurate picture drawn; $1\frac{1}{12}$
   b. Accurate picture drawn; $1\frac{5}{12}$
   c. Accurate picture drawn; $\frac{14}{15}$
   d. Accurate picture drawn; $1\frac{1}{3}$
   e. Accurate picture drawn; $1\frac{1}{2}$
   f. Accurate picture drawn; $1\frac{19}{21}$

2. Accurate picture and number sentence shown; $1\frac{5}{12}$ L

3. Accurate picture and number sentence shown; $1\frac{3}{8}$ tanks

Lesson 5

Sprint

Side A

1. $3\frac{1}{2}$
2. $2\frac{1}{2}$
3. $1\frac{1}{2}$
4. $\frac{1}{2}$
5. $\frac{2}{3}$
6. $1\frac{2}{3}$
7. $3\frac{2}{3}$
8. $3\frac{1}{3}$
9. $1\frac{1}{3}$
10. $1\frac{3}{4}$
11. $1\frac{1}{4}$
12. $2\frac{1}{4}$
13. $2\frac{3}{4}$
14. $3\frac{1}{4}$
15. $1\frac{9}{10}$
16. $2\frac{1}{10}$
17. $1\frac{3}{10}$
18. $3\frac{7}{10}$
19. $2\frac{4}{5}$
20. $2\frac{3}{5}$
21. $2\frac{1}{5}$
22. $2\frac{2}{5}$
23. $2\frac{7}{8}$
24. $2\frac{5}{8}$
25. $2\frac{3}{8}$
26. $2\frac{1}{8}$
27. $1\frac{1}{8}$
28. $3\frac{6}{7}$
29. $2\frac{1}{7}$
30. $1\frac{4}{7}$
31. $3\frac{3}{7}$
32. $2\frac{2}{7}$
33. $3\frac{1}{4}$
34. $1\frac{3}{8}$
35. $2\frac{7}{10}$
36. $3\frac{3}{5}$
37. $3\frac{4}{7}$
38. $2\frac{3}{10}$
39. $2\frac{1}{2}$
40. $3\frac{3}{4}$
41. $1\frac{1}{4}$
42. $3\frac{5}{6}$
43. $2\frac{2}{3}$
44. $1\frac{1}{3}$

Side B

1. $\frac{1}{2}$
2. $1\frac{1}{2}$
3. $2\frac{1}{2}$
4. $3\frac{1}{2}$
5. $\frac{3}{4}$
6. $1\frac{3}{4}$
7. $3\frac{3}{4}$
8. $3\frac{1}{4}$
9. $1\frac{1}{4}$
10. $1\frac{2}{3}$
11. $1\frac{1}{3}$
12. $2\frac{1}{3}$
13. $2\frac{2}{3}$
14. $3\frac{1}{3}$
15. $2\frac{9}{10}$
16. $1\frac{1}{10}$
17. $3\frac{3}{10}$
18. $2\frac{7}{10}$
19. $1\frac{4}{5}$
20. $1\frac{3}{5}$
21. $1\frac{1}{5}$
22. $2\frac{2}{5}$
23. $1\frac{7}{8}$
24. $1\frac{5}{8}$
25. $1\frac{3}{8}$
26. $1\frac{1}{8}$
27. $3\frac{1}{8}$
28. $2\frac{6}{7}$
29. $1\frac{1}{7}$
30. $3\frac{4}{7}$
31. $2\frac{3}{7}$
32. $1\frac{2}{7}$
33. $2\frac{1}{4}$
34. $3\frac{3}{8}$
35. $1\frac{7}{10}$
36. $2\frac{3}{5}$
37. $2\frac{4}{7}$
38. $1\frac{3}{10}$
39. $1\frac{1}{2}$
40. $2\frac{1}{4}$
41. $3\frac{3}{4}$
42. $2\frac{1}{6}$
43. $1\frac{1}{3}$
44. $3\frac{2}{3}$

Problem Set

1. a. Accurate picture drawn; $\frac{1}{12}$
   b. Accurate picture drawn; $\frac{1}{6}$
   c. Accurate picture drawn; $\frac{7}{12}$
   d. Accurate picture drawn; $\frac{11}{21}$
   e. Accurate picture drawn; $\frac{3}{8}$
   f. Accurate picture drawn; $\frac{13}{28}$
2. $\frac{7}{15}$ L
3. Answers will vary.

Exit Ticket

a. Accurate picture drawn; $\frac{5}{14}$

b. Accurate picture drawn; $\frac{1}{10}$

Homework

1. Pictures will vary; $\frac{5}{12}$
2. a. Accurate picture drawn; $\frac{1}{2}$
   b. Accurate picture drawn; $\frac{1}{6}$
   c. Accurate picture drawn; $\frac{7}{12}$
   d. Accurate picture drawn; $\frac{3}{10}$
   e. Accurate picture drawn; $\frac{4}{15}$
   f. Accurate picture drawn; $\frac{1}{21}$
3. $\frac{5}{8}$ lb
4. No; $\frac{6}{35}$ kg

Lesson 6

Problem Set

1. a. Accurate picture drawn; $\frac{11}{12}$
   b. Accurate picture drawn; $\frac{13}{15}$
   c. Accurate picture drawn; $\frac{7}{8}$
   d. Accurate picture drawn; $\frac{9}{10}$
   e. Accurate picture drawn; $\frac{20}{21}$
   f. Accurate picture drawn; $1\frac{1}{15}$
2. $\frac{5}{12}$ hr
3. Yes; answers will vary.

Exit Ticket

a. Accurate picture drawn; $\frac{7}{10}$
b. Accurate picture drawn; $\frac{1}{2}$

Homework

1. a. Accurate picture drawn; $\frac{1}{6}$
   b. Accurate picture drawn; $\frac{2}{3}$
   c. Accurate picture drawn; $\frac{13}{21}$
   d. Accurate picture drawn; $\frac{21}{40}$
   e. Accurate picture drawn; $\frac{13}{20}$
   f. Accurate picture drawn; $\frac{23}{24}$
   g. Accurate picture drawn; $\frac{15}{28}$
   h. Accurate picture drawn; $\frac{7}{12}$
2. $\frac{7}{8}$ m
3. $\frac{17}{24}$ kg

Lesson 7

Sprint

Side A

1. 1/2
2. 1/3
3. 1/4
4. 1/2
5. 1/3
6. 1/4
7. 1/2
8. 1/3
9. 1/4
10. 1/2
11. 1/3
12. 1/4
13. 2/3
14. 1/2
15. 1/3
16. 1/5
17. 2/3
18. 1/2
19. 1/3
20. 1/6
21. 2/3
22. 1/2
23. 1/3
24. 1/7
25. 2/3
26. 1/2
27. 1/3
28. 1/8
29. 2/3
30. 3/4
31. 3/4
32. 3/4
33. 5/6
34. 5/6
35. 4/5
36. 4/5
37. 4/5
38. 2/3
39. 3/4
40. 4/5
41. 5/6
42. 2/3
43. 5/6
44. 4/5

Side B

1. 1/2
2. 1/3
3. 1/4
4. 1/2
5. 1/3
6. 1/4
7. 1/2
8. 1/3
9. 1/4
10. 1/2
11. 1/3
12. 1/4
13. 2/3
14. 1/2
15. 1/3
16. 1/5
17. 2/3
18. 1/2
19. 1/3
20. 1/6
21. 2/3
22. 1/2
23. 1/3
24. 1/7
25. 2/3
26. 1/2
27. 1/3
28. 1/8
29. 2/3
30. 3/4
31. 3/4
32. 3/4
33. 4/5
34. 4/5
35. 4/5
36. 5/6
37. 5/6
38. 2/3
39. 3/4
40. 4/5
41. 5/6
42. 2/3
43. 4/5
44. 5/6

Problem Set

1. $\frac{2}{15}$
2. $\frac{1}{24}$
3. $\frac{9}{15}$ oz
4. Jim; $\frac{1}{12}$ gal
5. $\frac{1}{8}$

Exit Ticket

Mr. Pham; $\frac{13}{28}$

Homework

1. $\frac{1}{2}$
2. $\frac{2}{21}$
3. $\frac{7}{10}$ kg
4. $\frac{1}{4}$
5. $1\frac{1}{15}$ m

Lesson 8

Problem Set

1. a. $3\frac{1}{5}$
   b. $\frac{5}{8}$
   c. 8
   d. $1\frac{5}{7}$
   e. $17\frac{3}{4}$
   f. $1\frac{1}{3}$
   g. $32\frac{2}{3}$
   h. $79\frac{1}{8}$

2. $6\frac{2}{3}$ min

3. $6\frac{1}{2}$ hr

4. Answers will vary.

Exit Ticket

a. $6\frac{7}{8}$

b. $1\frac{1}{4}$

c. $11\frac{3}{8}$

d. $1\frac{4}{7}$

Homework

1. a. $4\frac{1}{4}$
   b. $\frac{3}{8}$
   c. 8
   d. $1\frac{2}{7}$
   e. $15\frac{4}{5}$
   f. $2\frac{1}{4}$
   g. $34\frac{5}{6}$
   h. $49\frac{5}{8}$

2. $5\frac{3}{8}$ m

3. $5\frac{1}{2}$ mi

4. Answers will vary.

Lesson 9

Sprint

Side A

1.	2/5	12.	4/4	23.	3/9	34.	9/12
2.	6/10	13.	2/12	24.	5/9	35.	6/8
3.	8/10	14.	2/3	25.	0/9	36.	5/12
4.	4/5	15.	1/12	26.	4/4	37.	4/5
5.	1/10	16.	11/12	27.	6/8	38.	4/5
6.	2/5	17.	11/12	28.	11/12	39.	3/6
7.	6/10	18.	3/6	29.	7/9	40.	2/12
8.	3/5	19.	3/6	30.	3/10	41.	5/15
9.	2/4	20.	3/6	31.	1/5	42.	4/15
10.	3/4	21.	3/3	32.	3/6	43.	9/15
11.	1/12	22.	3/12	33.	7/12	44.	6/15

Side B

1.	2/2	12.	3/3	23.	9/12	34.	3/3
2.	3/8	13.	2/9	24.	10/12	35.	3/4
3.	5/8	14.	2/9	25.	0/12	36.	5/12
4.	1/12	15.	4/9	26.	7/10	37.	3/5
5.	7/12	16.	4/12	27.	8/10	38.	0/5
6.	7/8	17.	1/12	28.	6/6	39.	4/12
7.	1/8	18.	3/12	29.	9/12	40.	8/8
8.	6/8	19.	2/10	30.	10/10	41.	6/8
9.	2/4	20.	6/8	31.	6/10	42.	1/10
10.	0/6	21.	3/10	32.	2/6	43.	9/10
11.	6/9	22.	1/10	33.	1/12	44.	4/12

Problem Set

1. a. $\frac{25}{28}$
   b. $1\frac{3}{8}$
   c. $\frac{45}{56}$
   d. $1\frac{1}{63}$
   e. $\frac{13}{15}$
   f. $1\frac{7}{12}$
   g. $\frac{25}{33}$
   h. $1\frac{17}{20}$

2. Answers will vary.

3. $1\frac{5}{8}$ gal

4. $1\frac{13}{20}$ kg; $\frac{17}{20}$ kg

Exit Ticket

a. $\frac{22}{24}$ or $\frac{11}{12}$

b. $1\frac{9}{10}$

Homework

1. a. $\frac{14}{15}$
   b. $\frac{38}{55}$
   c. $1\frac{1}{18}$
   d. $\frac{3}{4}$
   e. $1\frac{11}{15}$
   f. $1\frac{5}{24}$
   g. $2\frac{1}{12}$
   h. $2\frac{1}{12}$

2. $1\frac{5}{12}$ hours

3. $2\frac{17}{20}$ kg

4. $\frac{9}{40}$

Lesson 10

Sprint

Side A

1. 4
2. $3\frac{1}{2}$
3. $4\frac{1}{2}$
4. 2
5. $2\frac{1}{2}$
6. 2
7. $2\frac{1}{2}$
8. 3
9. $3\frac{1}{3}$
10. $3\frac{2}{3}$
11. $7\frac{2}{3}$
12. 8
13. $6\frac{3}{4}$
14. $8\frac{3}{4}$
15. $4\frac{3}{4}$
16. $3\frac{3}{4}$
17. $2\frac{3}{4}$
18. $\frac{3}{4}$
19. 6
20. $5\frac{5}{6}$
21. $5\frac{1}{6}$
22. $9\frac{5}{6}$
23. $10\frac{5}{6}$
24. $10\frac{5}{6}$
25. $7\frac{5}{6}$
26. $3\frac{5}{6}$
27. $5\frac{4}{5}$
28. $9\frac{7}{8}$
29. $5\frac{4}{5}$
30. $4\frac{5}{12}$
31. $9\frac{2}{5}$
32. $9\frac{2}{5}$
33. $9\frac{4}{5}$
34. $9\frac{4}{5}$
35. $7\frac{2}{3}$
36. $7\frac{2}{3}$
37. $5\frac{2}{3}$
38. $5\frac{1}{3}$
39. 5
40. $4\frac{6}{7}$
41. $4\frac{6}{7}$
42. $5\frac{1}{5}$
43. $5\frac{2}{5}$
44. $6\frac{2}{9}$

Side B

1. 3
2. $2\frac{1}{2}$
3. $3\frac{1}{2}$
4. 1
5. $1\frac{1}{2}$
6. 3
7. $3\frac{1}{2}$
8. 4
9. $4\frac{1}{3}$
10. $4\frac{2}{3}$
11. $8\frac{2}{3}$
12. 9
13. $7\frac{3}{4}$
14. $9\frac{3}{4}$
15. $5\frac{3}{4}$
16. $4\frac{3}{4}$
17. $3\frac{3}{4}$
18. $\frac{3}{4}$
19. 7
20. $5\frac{5}{6}$
21. $5\frac{1}{6}$
22. $9\frac{5}{6}$
23. $10\frac{5}{6}$
24. $10\frac{5}{6}$
25. $6\frac{5}{6}$
26. $4\frac{5}{6}$
27. $6\frac{4}{5}$
28. $9\frac{7}{8}$
29. $4\frac{4}{5}$
30. $5\frac{5}{12}$
31. $8\frac{2}{5}$
32. $8\frac{2}{5}$
33. $8\frac{4}{5}$
34. $8\frac{4}{5}$
35. $6\frac{2}{3}$
36. $6\frac{2}{3}$
37. $6\frac{2}{3}$
38. $6\frac{1}{3}$
39. 6
40. $6\frac{6}{7}$
41. $6\frac{6}{7}$
42. $4\frac{1}{5}$
43. $4\frac{1}{5}$
44. $7\frac{2}{9}$

Problem Set

1. a. $3\frac{9}{20}$
   b. $4\frac{3}{20}$
   c. $3\frac{8}{15}$
   d. $6\frac{1}{15}$
   e. $8\frac{1}{21}$
   f. $8\frac{11}{21}$
   g. $18\frac{33}{40}$
   h. $21\frac{1}{40}$
2. $8\frac{1}{4}$ mi
3. $6\frac{1}{12}$ gal
4. No; answers will vary.

Exit Ticket

1. $4\frac{5}{6}$
2. $8\frac{13}{28}$

Homework

1. a. $3\frac{7}{10}$
   b. $4\frac{1}{10}$
   c. $4\frac{8}{15}$
   d. $5\frac{4}{15}$
   e. $6\frac{19}{21}$
   f. $8\frac{8}{21}$
   g. $19\frac{23}{40}$
   h. $20\frac{31}{40}$
2. $8\frac{1}{2}$ hr
3. $6\frac{1}{12}$ m
4. Answers will vary.

Lesson 11

Problem Set

1. a. $\frac{1}{6}$
   b. $\frac{11}{30}$
   c. $\frac{1}{8}$
   d. $1\frac{1}{40}$
   e. $1\frac{2}{15}$
   f. $1\frac{2}{15}$
   g. $3\frac{4}{21}$
   h. Number lines drawn
2. 24 shown as denominator; answers will vary.
3. $\frac{11}{12}$ L; $\frac{1}{4}$ L
4. $2\frac{1}{14}$ kg

Exit Ticket

a. $\frac{9}{20}$

b. $2\frac{1}{6}$

Homework

1. a. $\frac{3}{10}$
   b. $\frac{13}{24}$
   c. $\frac{1}{10}$
   d. $1\frac{1}{6}$
   e. $1\frac{1}{20}$
   f. $2\frac{4}{21}$
   g. $10\frac{1}{8}$
   h. $12\frac{7}{24}$
2. $\frac{14}{24}$ or $\frac{7}{12}$
3. $2\frac{3}{14}$ yd
4. $\frac{2}{15}$
5. $3\frac{1}{6}$ gal

Lesson 12

Sprint

Side A

1.	1/4	12.	3/10	23.	1/10	34.	1/12
2.	1/4	13.	3/10	24.	1/12	35.	1/12
3.	1/6	14.	3/10	25.	1/12	36.	2/15
4.	1/6	15.	1/10	26.	5/12	37.	4/15
5.	1/8	16.	3/10	27.	5/12	38.	7/15
6.	1/8	17.	3/10	28.	1/12	39.	2/15
7.	5/8	18.	1/10	29.	5/12	40.	7/12
8.	5/8	19.	7/10	30.	5/12	41.	7/16
9.	3/8	20.	7/10	31.	1/12	42.	7/15
10.	3/10	21.	3/10	32.	7/12	43.	5/12
11.	3/10	22.	3/10	33.	5/12	44.	5/12

Side B

1.	1/10	12.	3/10	23.	3/8	34.	1/12
2.	1/10	13.	1/10	24.	4/15	35.	1/12
3.	1/4	14.	7/10	25.	4/15	36.	1/12
4.	1/4	15.	7/10	26.	2/15	37.	2/15
5.	3/10	16.	3/10	27.	2/15	38.	8/15
6.	3/10	17.	3/10	28.	1/15	39.	4/15
7.	1/10	18.	1/10	29.	8/15	40.	5/12
8.	3/10	19.	1/8	30.	3/15	41.	5/16
9.	3/10	20.	1/8	31.	1/12	42.	8/15
10.	1/10	21.	5/8	32.	7/12	43.	7/12
11.	3/10	22.	7/8	33.	5/12	44.	7/16

Problem Set

1. a. $\frac{19}{20}$
   b. $\frac{13}{20}$
   c. $2\frac{13}{15}$
   d. $1\frac{11}{15}$
   e. $\frac{20}{21}$
   f. $6\frac{17}{21}$
   g. $11\frac{5}{6}$
   h. $14\frac{23}{24}$

2. No; number line drawn correctly

3. $4\frac{17}{20}$ gal

4. $2\frac{7}{12}$ hr

Exit Ticket

1. $4\frac{1}{6}$

2. $2\frac{11}{12}$

Homework

1. a. $\frac{11}{12}$
   b. $\frac{11}{12}$
   c. $1\frac{19}{20}$
   d. $1\frac{17}{20}$
   e. $\frac{20}{21}$
   f. $4\frac{20}{21}$
   g. $12\frac{7}{8}$
   h. $14\frac{23}{40}$

2. Yes; number line drawn correctly

3. $4\frac{13}{20}$ gal

4. $2\frac{7}{12}$ ft

Lesson 13

Problem Set

1. a. Less than 1
   b. Greater than 1
   c. Less than 1
   d. Greater than 1
2. a. Greater than $\frac{1}{2}$
   b. Less than $\frac{1}{2}$
   c. Less than $\frac{1}{2}$
   d. Greater than $\frac{1}{2}$
3. a. $>$
   b. $<$
   c. $=$
   d. $<$
4. No; answers will vary.
5. No
6. No

Exit Ticket

1. a. Less than 1
   b. Greater than 1
   c. Less than $\frac{1}{2}$
   d. Greater than $\frac{1}{2}$
2. $<$

Homework

1. a. Less than 1
   b. Greater than 1
   c. Less than 1
   d. Less than 1
2. a. Less than $\frac{1}{2}$
   b. Greater than $\frac{1}{2}$
   c. Less than $\frac{1}{2}$
   d. Greater than $\frac{1}{2}$
3. a. $<$
   b. $<$
   c. $>$
   d. $<$
4. No; answers will vary.
5. No
6. Yes

Lesson 14

Sprint

Side A

1.	1/2	12.	1/4	23.	1/3	34.	5/6
2.	1/3	13.	2/3	24.	1/7	35.	4/5
3.	1/4	14.	1/2	25.	2/3	36.	4/5
4.	1/2	15.	1/3	26.	1/2	37.	4/5
5.	1/3	16.	1/5	27.	1/3	38.	2/3
6.	1/4	17.	2/3	28.	1/8	39.	3/4
7.	1/2	18.	1/2	29.	2/3	40.	4/5
8.	1/3	19.	1/3	30.	3/4	41.	5/6
9.	1/4	20.	1/6	31.	3/4	42.	2/3
10.	1/2	21.	2/3	32.	3/4	43.	5/6
11.	1/3	22.	1/2	33.	5/6	44.	4/5

Side B

1.	1/2	12.	1/4	23.	1/3	34.	4/5
2.	1/3	13.	2/3	24.	1/7	35.	4/5
3.	1/4	14.	1/2	25.	2/3	36.	5/6
4.	1/2	15.	1/3	26.	1/2	37.	5/6
5.	1/3	16.	1/5	27.	1/3	38.	2/3
6.	1/4	17.	2/3	28.	1/8	39.	3/4
7.	1/2	18.	1/2	29.	2/3	40.	4/5
8.	1/3	19.	1/3	30.	3/4	41.	5/6
9.	1/4	20.	1/6	31.	3/4	42.	2/3
10.	1/2	21.	2/3	32.	3/4	43.	4/5
11.	1/3	22.	1/2	33.	4/5	44.	5/6

Problem Set

1. a. 5
   b. $2\frac{1}{4}$
   c. 1
   d. 0

2. a. $4\frac{1}{15}$
   b. $\frac{3}{40}$
   c. 1
   d. $58\frac{11}{12}$
   e. $3\frac{2}{35}$
   f. 4.3 or $4\frac{3}{10}$

3. $31\frac{1}{8}$ lb

4. $18\frac{9}{20}$ kg or 18.45 kg

Exit Ticket

1. $5\frac{7}{12}$

2. $5\frac{1}{30}$

Homework

1. a. 3
   b. $3\frac{1}{4}$
   c. $1\frac{6}{7}$
   d. 0

2. a. $4\frac{27}{28}$
   b. $2\frac{11}{12}$
   c. 1
   d. $37\frac{7}{8}$
   e. $2\frac{19}{30}$
   f. $18\frac{2}{10}$ or 18.2

3. $9\frac{1}{6}$ yd

4. $3\frac{19}{36}$ lb

Lesson 15

Sprint

Side A

1.	1/4	12.	3/10	23.	1/8	34.	5/12
2.	1/2	13.	3/10	24.	1/4	35.	2/3
3.	1/2	14.	3/5	25.	1/4	36.	7/12
4.	1/2	15.	1/10	26.	1/4	37.	1/8
5.	1/10	16.	4/5	27.	1/6	38.	1/8
6.	3/10	17.	1/9	28.	1/6	39.	7/12
7.	5/12	18.	2/9	29.	1/6	40.	5/12
8.	1/2	19.	1/3	30.	2/3	41.	6/7
9.	1/2	20.	1/3	31.	2/9	42.	62/72
10.	1/5	21.	1/12	32.	4/9	43.	8/9
11.	1/10	22.	1/3	33.	1/12	44.	11/12

Side B

1.	1/6	12.	2/5	23.	1/4	34.	5/12
2.	1/2	13.	1/10	24.	1/4	35.	3/4
3.	1/8	14.	3/5	25.	1/12	36.	7/12
4.	3/8	15.	3/10	26.	1/4	37.	1/12
5.	1/2	16.	7/10	27.	1/12	38.	5/6
6.	1/2	17.	1/6	28.	1/6	39.	5/8
7.	1/12	18.	1/3	29.	1/9	40.	3/8
8.	1/2	19.	1/3	30.	2/3	41.	34/42
9.	1/10	20.	1/3	31.	5/9	42.	7/8
10.	1/5	21.	1/3	32.	2/3	43.	47/54
11.	2/5	22.	1/3	33.	1/2	44.	65/72

Problem Set

1. 31 min 35 sec
2. $4\frac{9}{20}$ kg
3. $\frac{1}{40}$ oz
4. 6 min 27 sec
5. $\frac{11}{20}$ min

Exit Ticket

$1\frac{9}{10}$ dollars; $1.90

Homework

1. $\frac{7}{12}$ lb
2. $1\frac{4}{5}$ kg
3. $28\frac{2}{3}$ mi
4. $2\frac{11}{24}$ lb
5. $2\frac{1}{3}$ hr

Lesson 16

Problem Set

1. a. Accurate drawing shown
   b. Accurate drawing shown
   c. Accurate drawing shown
   d. Accurate drawing shown
2. $1\frac{3}{7}$ or $\frac{10}{7}$ ft longer
3. $1\frac{1}{7}$ or $\frac{8}{7}$ ft

Exit Ticket

a. Accurate drawing shown
b. Accurate drawing shown
c. Accurate drawing shown

Homework

a. Accurate drawing shown
b. Accurate drawing shown
c. Accurate drawing shown; $\frac{1}{10}$; 21 mi
d. Answers will vary.

This page intentionally left blank